THE GEOLOGY OF SOUTHERN NEW MEXICO

A Beginner's Guide

THE GEOLOGY OF SOUTHERN NEW MEXICO

A BEGINNER'S GUIDE

Including El Paso

GREG H. MACK

University of New Mexico Press
Albuquerque

© 1997 by the
University of New
Mexico Press
All rights reserved.
First Edition

Library of Congress
Cataloging-in-
Publication Data
Mack, Greg H.
The geology of southern
New Mexico: a
beginner's guide
Greg H. Mack.—1st ed.
  p.   cm.
Includes bibliographical
references (p.).
ISBN 0-8263-1794-4 (pbk.)
1. Geology—New Mexico.
2. Geology—Texas—El
  Paso Region.
I. Title.
QE143.M33   1977
557.899—dc20
96-35688

TO ANDY AND CORY,
AND MOST OF ALL
TO LELA

Contents

Prologue: What Is This Book About and Who Is It For? 1

Part I: The Earth Is Changing 3

1. How to Read the Rocks 5
2. Geologic Time 17
3. Deformation 24
4. Plate Tectonics 28

Part II: A Brief History of West Texas and Southern New Mexico Before Man Arrived 35

5. Precambrian: Collision and Accretion 39
6. Early and Middle Paleozoic: Shallow, Subtropical Seas 43
7. Late Paleozoic: The Ancestral Rocky Mountains 47
8. Early Cretaceous: The Chihuahua Trough 51
9. Late Cretaceous: The Western Interior Seaway 55
10. Early Tertiary: Laramide Mountain Building and Volcanism 59
11. Middle Tertiary: The Great Calderas 64
12. Late Tertiary and Quaternary: The Rio Grande Rift 67

Part III: Field Trips 73

1. Eastern Trans-Mountain Road, El Paso, TX 75
2. Cobre Uplift, near San Lorenzo, NM 79
3. McKelligon Canyon Park, El Paso, TX 82

4. Murchison Park, Scenic Drive, El Paso, TX   85
5. Bishop's Cap, near Las Cruces, NM   87
6. Lake Valley, NM   91
7. Dry Canyon–La Luz Canyon, Alamogordo, NM   93
8. Lucero Arroyo, Doña Ana Mountains, NM   98
9. Robledo Mountains, near Las Cruces, NM   103
10. Guadalupe Mountains, TX and NM   106
11. Cerro de Cristo Rey, El Paso, TX   112
12. Mescal Canyon, Truth or Consequences, NM   117
13. San Diego Mountain, NM   122
14. Faulkner Canyon, NM   125
15. Cedar Hills, NM   129
16. Florida Mountains, near Deming, NM   133
17. Dripping Springs State Park and San Augustin Pass, near Las Cruces   136
18. Black Range, NM   140
19. Apache Canyon, Caballo Mountains, NM   146
20. Rincon, NM   152
21. Grama, NM   156
22. Box Canyon, near Las Cruces, NM   158

Glossary of Terms   163

Further Reading   169

Index   170

Prologue

What Is This Book About and Who Is It For?

This book is designed for those interested in learning more about the history of the earth, especially as it affects the region of west Texas and southern New Mexico. It is for the curious and the imaginative; for those people who look at the rocks and the landscape around them and wonder how and when they formed. Why are some rocks red and others black? Why do some rocks appear to be tilted, whereas others are horizontal? How is it that fossils of animals that once lived in the sea can be found on top of some of the highest mountains? This book has answers to those and other questions.

This book is also for those who like to be outside. It is for those who enjoy viewing the natural world the way it should be viewed—in the field. It is for those who like to hike and climb, or just be outdoors on a warm sunny day.

This book was specifically written for beginners. It presumes no prior knowledge of geology, and no more than a rudimentary knowledge of any of the physical or biological sciences. As a consequence, part I of the book outlines in simple terms the fundamental principles of geology needed to interpret earth history in the field, including the origin of rocks, geologic time, rock deformation, and plate tectonics. This introductory material lays the groundwork for part II, which is a more detailed look at the geologic history of west Texas and southern New Mexico. What you learn in the first two parts of the book is then incorporated into field work via a series of field trips: places you can go to see the great variety of geology in our region. The field trips have detailed road logs, so that you can easily find your way to the outcrops, and descriptions and discussions of the rocks that make up the outcrops. Some trips are within a short distance of the cities of El Paso, Las Cruces, and other towns in the region, and some are in more isolated places. Most of the outcrops are easily accessible, indeed most are roadcuts, but some require hikes or climbs of varying degree of difficulty. In short, the trips are selected to accommodate the budgets and physical abilities of as many people as possible.

The equipment required for the field geologist is minimal. You need only a hammer, preferably a rock pick or mason's hammer, a hand lens with a 10 power lens, a notebook and pen or pencil to record your observations, and this book. Of course, caution should always be exercised in the deserts and mountains of the Southwest. Take lots of water and wear a hat. Sunblock is also a good idea. If you hike any significant distance from the road, it is best to have one or more partners in the off chance an accident occurs. Hiking boots are best, but tennis shoes will do for most locations. Cowboy boots are unreliable footwear in the field. If you've ever walked over rough ground in cowboy boots, you will quickly realize why cowboys ride horses.

I have tried to select safe outcrops, but there are always potential dangers. Avoid climbing on loose rock or standing beneath unstable overhangs. If you choose to hammer on a rock, it may be best to wear safety glasses; also, never use the hammer when your partner is close by. Finally, be aware of creatures living under the rocks. Snakes and scorpions like rocks almost as much as I do. Although these warnings may seem dire, the desert can be a safe place if you are careful. Just use common sense, and enjoy the rocks!

PART I

The Earth Is Changing

The earth is a dynamic place, constantly changing. The landscape we see today is quite different from landscapes of the geologic past. Where mountains now exist, there were in the past flat, tree-covered plains or warm, tropical seas. The roots of extinct volcanoes as large as Mount St. Helens can be found in inconspicuous hills and mesas scattered across the countryside. Where cattle now graze on the desert scrub, dinosaurs once roamed across green forests of pine, palm, and hardwoods. All of this rich history is there for those who are curious enough and willing to invest the time necessary to learn to read the rocks.

Geology, like any science, can be approached at a variety of different levels of complexity. The goal in this book is to provide the reader with a basic understanding of geologic processes and the ability to recognize geologic features in the field. The fundamental concepts presented in part I are the identification and origin of rocks (chapter 1), geologic time (chapter 2), rock deformation (chapter 3), and plate tectonics (chapter 4).

I have attempted to limit the geologic jargon as much as possible, although there are still many terms unique to geology that cannot be avoided. In order to help the reader with the specialized terminology, most of the key words in the following text are italicized and appear in a Glossary at the end of the book. The unfamiliar geologic terms may be daunting at first, but as a result of repeated usage throughout the book and application in the field, their meaning will become clear.

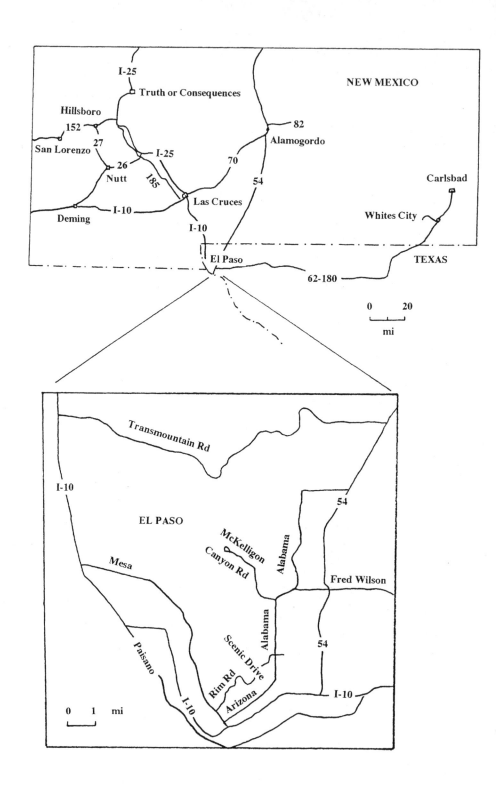

1

How to Read the Rocks

Solid matter on the earth, moon, and other celestial bodies in our solar system is composed of one or a combination of a hundred or so *elements*, such as carbon, hydrogen, oxygen, lead, and so on. These elements are the symbols that appear in chemistry classrooms on those large wall charts known as the periodic table, which all good physical scientists have either memorized or carry as a small, plastic card in their wallets. Don't worry, though; you do not need to learn the periodic table of elements or even buy the wallet-sized version to understand the fundamentals of geology in the field. I mention elements only because they are the building blocks of the earth and must be considered if we wish to start at the beginning, which we do.

It is rare in nature for an element to exist by itself, although some—the inert gases, helium and argon for example—are only found that way. It is more common for elements to combine together to make something more complex. Curiously, the combination of elements has properties completely different from those of the individual elements that make up the combination. Sodium and chlorine combine to make sodium chloride, also known as table salt. The salt is unlike either elemental sodium or chlorine; it is a unique substance. The natural combinations of elements in the world of geology are *minerals*. You may have heard of some common minerals: quartz, which vibrates in your wristwatch; or calcite, which makes up a clam shell or a chicken egg; or the mineral pyrite, also known as fool's gold. Despite the fact that thousands of different types of minerals exist on earth, only a handful of them are common. To be a geologist, it is necessary to be able to recognize the common minerals, and maybe even some of the uncommon ones, but at the level of this book we will only deal with a few easily identifiable minerals that are commonly encountered on the field trips.

Just as it is uncommon in nature for elements to exist in isolation, so too is it rare for a mineral to exist by itself. It is much more common for minerals to be found in natural combinations, with two or three or more minerals mixed together. These natural combinations of minerals are called *rocks*, the focus of this chapter. Rocks fall into three major categories: igneous, sedimentary, and metamorphic, all of which are found in our region and can be seen on the field trips.

Igneous Rocks.

An *igneous rock* begins its life as molten magma, which is generated anywhere from five to sixty miles below the earth's surface. Because hot, liquid magma is less dense than the surrounding solid rock, it rises toward the surface. As the magma rises it cools, causing minerals to grow, and ultimately produces that mix of minerals called an igneous rock.

We are most familiar with magma that reaches the earth's surface and forms *volcanic rocks*. Some volcanoes, like those in Hawaii, erupt passively as rivers of molten lava that flow easily down the flanks of the volcano (Figure 1). Many solidified lava flows preserve a ropy or wavy structure on the flow top that can be easily seen in the field. Many lava flows also have small holes or *vesicles* that develop as gases escape the lava. Other volcanoes erupt violently and present a significant danger to people who make their homes near them. Mount St. Helens in Washington State or Mount Pinatubo in the Philippines are good examples. These volcanoes differ from those that commonly erupt as lava flows by having a more viscous or thicker lava with more dissolved gas. Because the gas has trouble escaping through the gooey lava, it builds up and eventually explodes. The violent eruption creates a thick cloud of superheated gas mixed with fragments of rock, mineral, and glass. Collectively known as *pyroclastic* material, from the Greek "pyro" meaning "fire" and "clastic" meaning "particles," the fragments range in size from a house to dust. Volcanic rocks composed of pyroclastic material are called *tuffs*. *Volcanic glass*, a common component of pyroclastic material, but also originating as lava flows, represents lava that cools so quickly in

Figure 1. Common processes responsible for formation of volcanic rocks.

contact with air that minerals do not have the chance to form. Most of the volcanic glass in tuffs exists as angular microscopic pieces known as *ash*, or as larger pieces, some the size of boulders, of *pumice*. Pumice is a honeycomb of glass and vesicles and can be thought of as instantly solidified froth, like freezing the head of a beer. Pumice floats on water, because it is mostly air-filled vesicles.

The cloud of gas and volcanic particles emanating from an explosive volcanic eruption often rises tens of thousands of feet into the air, before the fragments of rock, mineral, and glass begin to settle out over the countryside (Figure 1). The finest material, the ash, can be carried hundreds to thousands of miles away from the volcano by the wind. Near Las Cruces, two separate ash deposits, each about a foot thick, can be tied to far-off volcanic eruptions (see Field Trip 21). One ash was derived from an eruption about 750,000 years ago in eastern California, one thousand miles away, and the other was transported thirteen hundred miles from a 600,000-year-old eruption at Yellowstone National Park.

In some instances, the cloud of gas and volcanic particles is heavier than air and sinks to the ground upon leaving the volcanic vent. These superheated clouds, known as *ash flows*, rush down the flanks of the volcano at speeds of hundreds of miles an hour, incinerating anything in their paths and ultimately depositing a thick layer tens to hundreds of feet thick of pumice-rich volcanic tuff (Figure 1). Ash flows are the most fearsome consequence of explosive eruptions, and were responsible for the destruction of Pompeii and Herculaneum in Italy in the first century A.D.

Another consequence of explosive eruptions is collapse of the top of the volcano, producing a circular depression, sometimes tens of miles in diameter, called a *caldera*. The state of New Mexico has one of the best preserved and most accessible calderas in North America. The *Valles* caldera, located just west of the city of Los Alamos, is about fifteen miles in diameter and formed during an explosive eruption only one million years ago—young in geologic terms. The ash flows that poured out of the *Valles* caldera are well exposed at Bandelier National Monument. Pumice from several eruptions of the *Valles* volcano was carried downriver by the ancestral Rio Grande and deposited near Las Cruces.

There is yet a third type of deposition of volcanic rock other than lava flows and pyroclastic deposits. A volcanic mudflow or *lahar* forms when heavy rain or snowmelt mixes with ash and other pyroclastic debris and produces a slurry that gravity pulls downslope (Figure 1). A lahar may develop during an eruption or between eruptions and is capable of burying villages tens of miles from the volcano. A lahar is easily recognized by the chaotic mixture of volcanic rock of all sizes, from ash to generally well-rounded boulders.

One final feature of volcanic rocks is worth mentioning at this time, because it is easily seen in outcrop, both up close and from considerable distances. When lava flows and ash flows cool and solidify the rock shrinks, resulting in vertically oriented cracks. In some cases the cracks form symmetrical columns called *columnar joints*, whereas other, less distinct cracks are referred to as cooling cracks.

The other major type of igneous rock represents magma that solidifies many miles below the earth's surface. This type of igneous rock constitutes the *plutonic rocks*, named after Pluto, the Greek god of the underworld. There are many places in west Texas and southern New Mexico, and elsewhere in the world, where uplift and erosion has exposed ancient plutonic igneous rocks. Although more difficult to understand than volcanic rocks, because we cannot observe their formation, it is possible to simulate the origin of plutonic rocks in the laboratory and through computer models. When exposed at the earth's surface plutonic rocks commonly undergo the process of *spheroidal weathering*, resulting in a very distinctive landscape of knobby hills resembling haystacks.

Igneous rocks are given names based on whether they are volcanic or plutonic and on the composition of the magma from which the rock solidified. These names are shown in the classification scheme in Figure 2. Volcanic rocks can be distinguished from plutonic rocks by the size of the minerals that make up the rock. Because volcanic rocks cool quickly, the minerals are very small, usually too small to be seen with the naked eye or with a hand lens, although a few scattered coarse minerals sometimes are present. In contrast, the slow-cooling plutonic rocks have larger minerals easily discernible with the unaided eye. The compositional groups can be distinguished

size of minerals \ color	light (white, tan, red, pink)	intermediate (gray, purple, green)	dark (black, dark green)
fine (volcanic)	rhyolite (tuff)	andesite (tuff)	basalt (tuff)
coarse (plutonic)	granite	diorite	gabbro

Figure 2. Classification of igneous rocks. The term "tuff" is added to the rock name for pyroclastic volcanic rocks.

by the color of the igneous rock. For example, the classification chart of Figure 2 shows that the dark-colored igneous rocks are called *basalt*, if volcanic, and *gabbro*, if plutonic. These rocks are dark because the most common minerals in them are dark. On the other end of the spectrum are the light-colored igneous rocks, the volcanic *rhyolite* and the plutonic *granite*. Granites and rhyolites are light colored, either white, pink, or reddish, because they contain minerals of this color. Not surprisingly, the igneous rocks of intermediate composition are a mixture of light and dark minerals, resulting in colors of gray or purplish gray. These intermediate rocks are the volcanic *andesite* and the plutonic *diorite*. So it is very easy to classify igneous rocks in the field. Simply determine if the rock has coarse minerals easily seen with the unaided eye, making it a plutonic rock, or is composed primarily of minerals too small to be seen by eye or with a hand lens, a volcanic rock. Then use the color of the rock: dark plutonic=gabbro, gray plutonic=diorite, light plutonic=granite; dark volcanic=basalt, gray volcanic=andesite, light volcanic=rhyolite. What could be easier! Nature was kind to us when it made igneous rocks.

Sedimentary Rocks.

The second major type of rocks is *sedimentary rocks*, which form by a variety of processes at the earth's surface. The origin of sedimentary rocks is easy to understand, because we can observe sediment being transported and deposited all around us in rivers, in lakes, and on the beach. Even after millions of years of burial, solidification, and uplift, sedimentary rocks still closely resemble sediment being deposited in modern environments.

The material that makes up sedimentary rock originates by the destruction of other rocks at the earth's surface by *weathering*. A rock exposed at the earth's surface is attacked by water and ice, plants, and animals, causing the rock to disintegrate. Much of the original rock is broken into pieces that are eventually moved away and deposited elsewhere. It is also possible for chemical reactions during weathering to change some of the original minerals in the exposed rock to new, microscopic minerals called *clay* minerals, which can also be transported away from the site of weathering and deposited elsewhere. The larger particles and clay minerals produced by weathering are collectively called detritus and constitute a class of sedimentary rocks known as the *detrital sedimentary rocks*.

Clay minerals are only one consequence of chemical reactions during weathering, however. These reactions also release individual elements that are transported in solution by rivers and added to lakes and to the sea. These elements remain dissolved in the lake or seawater until some physical or biological process causes precipitation of the elements into minerals. The chemically precipitated minerals accumulate on the sea floor or lake bottom as the *chemical sedimentary rocks*.

Let us look more closely at the detrital sedimentary rocks. As gravity, running water, or wind move detritus away from the site of weathering it is segregated, or sorted, according to size. For example, wind cannot transport boulders, not even in New Mexico or west Texas in the springtime, but gravity-driven mudflows and floodwaters of mountain streams can. The consequence of sorting is that detrital sedimentary rocks fall into distinct size categories, shown on the classification diagram of Figure 3. The coarse, gravel-sized particles form a rock called a *conglomerate*; the rock composed of sand-sized detritus is a *sandstone*; the rock composed of silt-sized detritus is a *siltstone*; and the rock composed mostly of microscopic clay minerals should be called a claystone, but is more often called a *shale*.

At this stage, you may be asking yourself a very important question: how does loose sediment, like sand at the beach, turn into a solid rock like a sandstone? When loose sediment is buried it undergoes two largely simultaneous processes, compaction and cementation. Compaction squeezes the detritus together as a result of the weight of the overlying sediment. Compaction is often enough in itself to turn clayey sediment into a shale, but cementation is required to turn gravel, sand, and silt into conglomerate, sandstone, and siltstone. Cementation involves precipitation of minerals from groundwater that occupies the small spaces between the detrital par-

size of detritus	rock name
gravel	conglomerate
sand	sandstone
silt	siltstone
clay	shale

Figure 3. Classification of detrital sedimentary rocks.

composition	rock name
large calcite fossils	fossiliferous limestone
microscopic calcite	micrite limestone
sand-sized spheres of calcite	oolite limestone
dolomite	dolostone
gypsum or halite	evaporite

Figure 4. Classification of chemical sedimentary rocks.

ticles. The cement binds the detritus together, so that when it is once again exposed at the earth's surface it is a solid rock, not loose sediment.

Now let's follow the plight of the elements produced by weathering and transported in solution into lakes or the sea (Figure 4). The most common way to remove these elements from solution is for a living organism to do it in the process of making a shell or precipitating a mineral within its tissues. Most animals living in lakes or in the sea, such as snails or clams, which can live in either a lake or the sea, or corals, which only live in the sea, make shells of the mineral calcite. The shells accumulate on the lake bottom or sea floor, and are subsequently buried, compacted, and cemented, producing a rock called a *limestone*. In most limestones you can see the fossil shells, but in others the limestone is composed of microscopic crystals of calcite precipitated by algae. Under rare circumstances in highly agitated tropical seas, sand-sized particles of calcite, called *oolites*, can precipitate directly from the water, resulting in a separate type of limestone that does not involve the actions of living organisms. Unfortunately for geologists, limestones are easily altered. During burial some limestones are changed into a rock called a *dolomite* or *dolostone*, when the mineral calcite converts to the mineral dolomite. A common result of dolomitization is partial or complete loss of the original features of the limestone, making it difficult or impossible to interpret what type of limestone was present before dolomitization.

In some places on the earth high rates of evaporation can make lake water or seawater so concentrated in dissolved elements that they begin to spontaneously precipitate out as minerals that accumulate as sedimentary layers. These deposits, known as *evaporites*, include the minerals halite (table salt) and gypsum, a soft mineral used in plasterboard. Such a process is happening in the Great Salt Lake, Utah, Lake Lucero on White Sands Missile Range, and in the Red Sea.

Sedimentary rocks, whether detrital or chemical in origin, are originally deposited in horizontal layers. These layers are easily recognizable and are given the name of *beds* or *bedding*. As you drive around the region, the bedding of sedimentary rocks usually distinguishes them from igneous and metamorphic rocks.

When you see a sedimentary rock, you should ask the question, "in what environment was the sediment deposited?" Sedimentary environments are generally divided into nonmarine, that is, deposition on the land surface, and marine, deposition in the ocean. Marine environments are further divided into shallow and deep, with a depth of about six hundred feet, the depth of the edge of continental shelves, marking the boundary. In some cases the type of fossils in the sedimentary rock will be a key to the depositional environment. Many living organisms only live in shallow seas. If we see similar fossils in a limestone, sandstone, or shale, we can safely assume the sediment was deposited in a shallow marine environment. Another type of fossil, called a *trace fossil*, can also aid in the interpretation of depositional environment. Trace fossils include burrows, tracks, and trails made by animals living on or in the sediment, as well as traces of roots of plants growing in the sediment. Root traces obviously indicate the sediment is nonmarine in origin, or if shallow marine, was exposed to the air shortly after deposition. Moreover, as a general rule, marine sedimentary rocks have a much higher concentration of burrows, tracks, and trails than nonmarine rocks.

There are a few other common structures in sedimentary rocks that help in the interpretation of depositional environment (Figure 5). Running water and wind produce distinctive structures as they transport sand, silt, oolites, and fossil fragments. At low current velocities, the sedimentary particles are transported as ripples. These small, delicate structures survive compaction and cementation and are preserved in ancient sedimentary rocks as *ripple marks*. At higher velocities, the particles are transported as dunes, which stand a few feet to thirty feet high. Sediment is moved up the gentle, upcurrent side of the dune and then avalanches down the steep slope or falls from suspension onto the steep slope, producing an inclined layer. It is the inclined layers, known as *crossbeds*, that we see in the rock record as evidence of transportation and deposition as dunes. Although

Figure 5. Common sedimentary structures.
A. Ripple marks, formed by low-velocity transport of sand;
B. Crossbeds, formed by moderate-velocity transport of sand as dunes;
C. Trace fossils, mainly burrows.
D. Desiccation cracks.

it is perhaps easiest to associate dunes with wind transport, such as at White Sands National Monument, rivers and shallow marine currents also transport sediment as dunes. One final sedimentary structure that you may encounter is *desiccation cracks*, formed as sediment dries out. If you see desiccation cracks, you can be sure that the sediment was deposited in a nonmarine environment or on the shoreline.

Metamorphic Rocks.

The least common type of rock at the earth's surface, and probably the hardest to understand, are the *metamorphic rocks*. As the name implies, these rocks started out as another type of rock, either igneous or sedimentary, and were changed into new rocks. The change, or metamorphism, takes place below the earth's surface under conditions of high temperature

minerals/ nature of foliation	rock name
minerals too small to be seen with eye/ breaks into flat sheets	slate
shiny micas aligned in sheets	schist
light and dark mineral banding	gneiss
calcite or dolomite; possible micas or garnet; commonly lacks foliation	marble
quartz, possible feldspars and micas; may lack foliation	meta-quartzite

Figure 6. Classification of metamorphic rocks.

and pressure, but without involving melting. Like the plutonic igneous rocks, we cannot directly observe metamorphism taking place, but we can simulate it in the laboratory and through computer models.

During the process of metamorphism the original minerals of the parent rock recrystallize into new minerals, some of which are quite distinctive, such as garnet. If at the same time the rock is being squeezed together or pulled apart, the new minerals will be aligned, resulting in a metamorphic feature called *foliation*. Which metamorphic minerals form and the type of foliation depends both on the magnitude of the temperature and pressure and on the composition of the parent rock. Five of the most common metamorphic rocks are listed in Figure 6. *Slate*, which forms under conditions of low temperature and pressure, represents a metamorphosed shale. Foliation in the slate manifests itself as the tendency to break into flat sheets, indicating that the microscopic minerals in the slate are oriented in horizontal planes. Metamorphic rocks formed under higher conditions of temperature and pressure include *schist*, which is characterized by the presence of shiny minerals called micas aligned in a single plane, and *gneiss* (the "g" is silent), whose foliation is expressed as light and dark banding. Schist and gneiss can form from a variety of different types of parent rock. Other common metamorphic rocks are *marble*, which is a metamorphosed limestone or dolomite, and *metaquartzite*, which is metamorphosed sandstone. Most metamorphic rocks are created in the core of mountain ranges that form when continents collide, such as those in the Himalayas, Alps, and Appalachians.

2

Geologic Time

One of the most difficult concepts to grasp in geology is the vastness of geologic time. The earth is 4.6 billion years old; that is, 4,600,000,000 years. The length of a human life pales in comparison. In fact, all of human history is just the blink of an eye in the history of the earth. The landscape and lifeforms in west Texas and southern New Mexico have changed countless times in its geologic history on a time scale that we can scarcely imagine. That is not to say that geologic processes are imperceptible to us or beyond our comprehension. Volcanoes erupt, earthquakes lift the land surface, and summer floods deposit thin layers of sediment. Roman columns stand ten feet below the sea, which indicates that sea level has risen, the land has subsided, or both, in only a few thousand years. These are hints of the processes that shape the earth over its long history and provide a guide to the interpretation of the rock record.

In our everyday lives we deal with time in two ways, relative time and absolute time. Geology does the same. *Relative time* offers a comparison. For example, I am older than most of the students I teach. Although a true statement, this type of comparison is not very precise. It does not designate how old the students are or how old I am, nor does it give the difference in our ages. It merely states that I was here first. Relative time is easily applied to well-layered sedimentary or volcanic rocks, which were originally deposited in horizontal layers on the earth's surface. Obviously, the oldest rock is on the bottom of the pile, with progressively younger rocks above. Even if the rocks have been tilted by processes discussed in the next chapter, it is possible to tell which rock was deposited first, as long as the rocks are not upside down (yes, rocks can be completely overturned, but fortunately it is rare).

Relative time is harder to apply to metamorphic rocks, which have been heated and squeezed and rotated from their original position. I have looked at many metamorphic rocks and not known which was up! Determining the relative age of a plutonic igneous rock, such as a granite, with respect to the rocks that surround it can also be difficult. In many cases the molten magma moved upward through cracks and fissures in the overlying rock before crystallizing into a solid plutonic rock. The result is a *cross cutting relationship*, indicating that the plutonic rock is younger than the surrounding rock (Figure 7).

So when we examine an outcrop in the field, we can generally tell the relative age of the rocks. But what if we want to compare the age of rocks that are not in direct contact with each other, say, rocks in New Mexico with those in Montana, or New York, or Russia? The answer to this question did not become apparent until the beginning of the nineteenth century, when geologists learned that one way to compare rocks all over the world is by comparing the age of the fossils they contain. The types of organisms living on the earth have changed dramatically throughout geologic time. The assemblage of organisms alive 500 million years ago was much different than the life assemblage 400, or 300, or 50 million years ago. So if rocks contain the same unique assemblage of fossils, then they must have formed at about the same time; a pretty simple concept, but one that took many years of careful collection and description of fossils by geologists from all over the world before it was shown to be true.

Using fossils to compare the age of rocks resulted in the *geologic time scale*, one of the few facts in geology worth memorizing, although I carry a small plastic card in my wallet (Figure 8). The fundamental unit of time is the *Period*, which is based on nothing more than rocks whose fossils can be used as a global standard of comparison. The Period names come from places where the fossiliferous rocks were first described. For example, the Devonian Period is named for fossiliferous rocks in Devonshire, England; the Jurassic Period is named for rocks in the Jura Mountains, on the border of Switzerland and France; and the Mississippian Period is named for rocks exposed in Illinois—along the banks of the Mississippi River. It took over half a century for all of the Periods

Figure 7. Two examples of a cross cutting relationship, in which molten magma cut across previously existing rocks and then solidified into an igneous rock. This field relationship demonstrates that the igneous rock is younger than the rocks it cross cuts.
A. Light-colored granite cross cutting dark metamorphic rocks; granite is younger.
B. Steeply dipping dark dike of andesite cross cutting more gently dipping, lighter beds of andesite; darker andesite is younger.

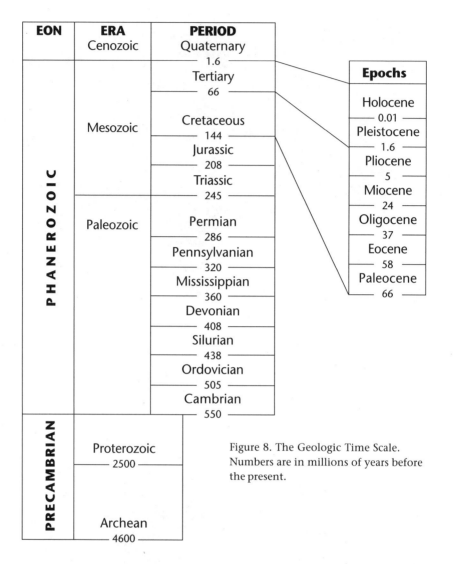

Figure 8. The Geologic Time Scale. Numbers are in millions of years before the present.

to be defined. Note also that there are five larger subdivisions of time called *Eras*, of which the Archean and Proterozoic have no Periods, and the Eras are grouped into two *Eons*, Precambrian and Phanerozoic. The Periods are divided into smaller units, called *Epochs*, but only the Epochs of the Tertiary and Quaternary Periods are shown in Figure 8. Why are there no Periods in the Archean and Proterozoic Eras? I'll get back to that in a moment.

There are several problems with using fossils to date rocks. One is that not all rocks have fossils and another is that not all organisms make fossils. Fossils are largely restricted to sedimentary rocks, although slate and volcanic ash sometimes have them too. Most igneous and metamorphic rocks cannot be dated by using fossils. Moreover, not all sedimentary rocks have fossils, the most common unfossiliferous rock being conglomerate. There is also a bias in the fossil record toward organisms with hard parts, such as shells, bones, and teeth. Soft-bodied organisms are much less likely to leave a fossil record, although a few excellent examples exist of carbon films or impressions of soft-bodied organisms. These two problems explain why there are no Periods in the Archean and Proterozoic Eras of the Precambrian Eon. During this time organisms did not have hard parts, and, because of their great age, most sedimentary rocks deposited in the Precambrian were subsequently metamorphosed or melted. The fossil record is simply not good enough to subdivide the Archean and Proterozoic into Periods. Application of the geologic time scale using fossils is primarily for sedimentary rocks of Cambrian or younger age. Although this may seem excessively restrictive, the bulk of rocks exposed at the earth's surface are Phanerozoic (Cambrian to Holocene) sedimentary rocks.

It is important to remember that the geologic time scale based on fossils represents relative geologic time. It is a simple comparison. If two rocks have the same unique fossils, we know that they were deposited at the same time, but we do not know when that was. Yet the geologic time scale shown in Figure 8 has absolute ages in millions of years before the present assigned to the boundaries between the Epochs, Periods, Eras, and Eons. Where do these numbers come from and how is it possible to go from relative time to absolute time?

Determining *absolute geologic time* is possible because of radioactive decay. A radioactive element changes to a new, stable element at a known, measurable rate, and the conversion is independent of other variables like heat and pressure. Radioactive decay is a clock ticking away in the rocks, waiting for us to read it. Unfortunately, it was not until the early part of the twentieth century, long after the geologic time scale based on fossils had taken shape, that we understood how the clock worked and had the sophisticated equipment required to read it.

The catch with radiometric dating is that it can only be applied to igneous and metamorphic rocks. It cannot be applied to sedimentary rocks because they are composed of minerals and elements derived from older rocks by the process of weathering. The minerals in a sandstone come from the breakdown of older rocks, so that if the mineral is dated by using radioactive decay the result is the age of the older rocks that supplied the mineral, not the time of deposition of the sand.

How then can we put absolute numbers on a time scale that is based on fossils in rocks that cannot be radiometrically dated? Once again nature has been kind to us. There are places on the earth where fossiliferous sedimentary rocks are interlayered with volcanic rocks. We can assign a geologic Period to the sedimentary rocks by using the fossils within them, while at the same time we can determine an absolute age of the volcanic rocks by radioactive decay. In the process, an absolute age for part of the geologic Period in question becomes available. It is also possible to date those igneous and metamorphic rocks of the Precambrian Eon and to accurately define the age of its boundaries. The result is a geologic time scale measured in absolute time, although it is constantly being refined as new data become available.

There is one final caveat to be considered in the concept of geologic time, the *unconformity*. A record of geologic events only exists for those places where rocks are forming—where sediment is being deposited, volcanoes are erupting, magma is crystallizing below the surface, or metamorphism is taking place. But what about a place where no rocks are forming or where rocks are being removed by erosion? In these places there will be a gap in the rock record called an unconformity. In most cases there is good evidence of the existence of an unconformity in the field (Figure 9). Sometimes there is clear evidence of erosion between sedimentary or volcanic layers. Another type of unconformity exists when a sedimentary rock directly overlies a plutonic igneous rock, like granite, or a metamorphic rock. The lost record in this case is the time it took for the plutonic or metamorphic rock, which formed below the earth's surface, to move

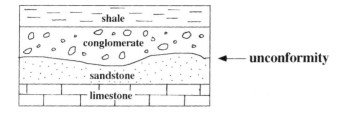

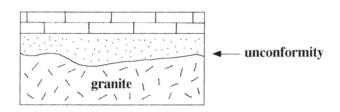

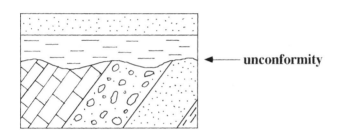

Figure 9. Types of unconformities in cross-sectional view.

upward and to be uncovered by erosion before deposition of the overlying rocks. Yet a third type of unconformity is when rocks with different angle of tilt are juxtaposed. The gap in the rock record is the time it took for the lower rocks to be tilted and eroded before deposition of the overlying layers. Other unconformities are much more subtle and can only be recognized by careful dating of the rocks, revealing a significant time gap.

So there is geologic time. From now on it is not enough to simply describe the type of rock. We need to know the age as well, even if it is just relative age.

3

Deformation

As you drive around the countryside of west Texas and southern New Mexico, you may observe that the layers of sedimentary and volcanic rocks in some of the mountain ranges are not horizontal. Instead, they are tilted at angles of 20, 30, 40 degrees or more. How is it that these rocks, which were deposited as horizontal layers, have been rotated from their original orientation to the angles we see today? You may also notice in your travels that there are places where granite and metamorphic rocks are exposed at the earth's surface. By what process have these rocks, which originally formed miles below the earth's surface, been moved upward to their present position?

The process responsible for the observations discussed in the preceding paragraph is *deformation*, the study of which constitutes a major subdiscipline within the science of geology called *structural geology*. Stresses within the earth cause rocks to deform by bending or breaking. Geologists use the term *dip* to refer to the direction and angle of tilt of rocks and the term *fault* to describe a break or fracture along which rocks on either side of the fracture move past one another. As we drive through the countryside we see the direct evidence of deformation in dipping beds and plutonic igneous and metamorphic rocks brought to the surface along faults.

The earth experiences two kinds of stress: *compression*, which is squeezing together, and *extension*, which is pulling apart. The magnitude of stress varies from place to place in the earth, as well as varying through time. The region of west Texas and southern New Mexico is currently under an extensional stress, but there have been several periods of compression in the geologic past. In some places the stress is too weak to result in rock deformation, but in our region the extensional stress has been strong enough over the past ten million years or so to create the mountains and basins that exist today.

Normal Fault

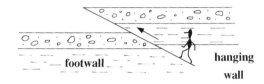

Thrust Fault

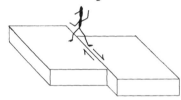

Strike-Slip Fault

Figure 10. Types of faults.

The application of compressional or extensional stress causes strain to build up in the earth. Once the strain exceeds a certain critical limit, the rocks break, creating a fault. As the fault moves it releases energy in the form of earthquake waves, which pulsate through the earth like ripples on a pond when a rock is thrown in it.

There are three types of faults, shown in Figure 10. A fault produced by extension, or pulling apart, is a *normal fault*, so named because it was the first and most common type of fault recognized by early miners and geologists. The miners also named the two sides of the fault. Mine shafts often were dug along the fault surface itself, because that is where the ore minerals crystallized. So each working day the miners walked down the fault surface toward the active part of the mine. They called the side of the fault they walked on the *footwall* and the side above them, where they hung their lanterns, the *hanging wall*. These terms have persisted in the geologic lexicon and are very useful in describing faults. For a normal fault, the footwall has moved up relative to the hanging wall. Most of the mountain ranges in west Texas and southern New Mexico are bordered by normal faults, with the mountains constituting the footwall and the adjacent basin the hanging wall. It is also common during its history for a normal fault to rotate, which causes rocks in the footwall to dip away from the fault. This is readily apparent in the Franklin Mountains near El Paso, where the main border fault is on the east side of the range and the rocks in the range dip to the west.

Compressive stress results in a different type of fault, which probably should be called an "abnormal" fault, but is actually called a *thrust fault* (Figure 10). In a thrust fault the hanging wall is up and the footwall is down. Compression also produces deformational structures known as folds, which represent bending or buckling of rocks without breakage. The upturned fold is an *anticline* and the downturned fold is a *syncline*. Folds and thrust faults are present in west Texas and southern New Mexico, but are much less common and more difficult to access than normal faults. An exception is the presence of large folds in Sierra Juárez, which are easily visible from various vantage points in El Paso. Although rare in our region, folds and thrust faults are the norm in the Appalachian and Ouachita Mountains and in the Rocky Mountains of Colorado, Wyoming, Utah, Montana, and Canada.

The third type of fault exists where rocks on opposite sides of the fault slide past each other with no vertical displacement (Figure 10). This fault has the unusual name of *strike-slip fault*. The best example of a strike-slip fault is the San Andreas fault in California, which is really a series of interconnected faults, a few of which, like the fault that generated the January 1994 Northridge earthquake, are thrust faults. The fault responsible for the Northridge earthquake is also noteworthy as an example of a fault that does not reach the surface. Strike-slip faults are of only minor importance in the geologic history of west Texas and southern New Mexico.

Let's consider in more detail how faulting actually works. Fault movement, especially that of normal and strike-slip faults, is episodic. Strain builds up for hundreds or thousands of years before it is released by offset of a few feet or tens of feet on a fault. Before the fault can move again, the strain must build up to the critical level. If the recurrence interval, that is, the time between fault movement, is long in human terms, then we tend to view the fault as being inactive or extinct. That appears to be the case in west Texas and southern New Mexico, where there has not been a major fault movement and associated earthquake in recent history. There is ample evidence, however, that a large earthquake, perhaps in the range of 6 or 7 on the Richter scale, is possible, although certainly less likely than in California. Our region is riddled with geologically young faults, a few of which can be shown to have moved tens of feet sometime in the past few thousand years. Most notable among these are the Organ Mountains fault, located just west of White Sands Missile Range headquarters, and the Caballo fault, positioned between Elephant Butte and Caballo dams. Other evidence of strain buildup and release are small earthquakes, most detectable only by sensitive seismological equipment, and the 1995 earthquake near Alpine, Texas, which measured around 5 on the Richter scale. Although seemingly far away, Alpine is located within the same belt of crustal extension and normal faulting as El Paso and southern New Mexico. Before you go screaming into the street or put your house up for sale, let me reemphasize that the recurrence interval on the faults in our region appears to be quite long, probably at least a thousand years or more. Although the chance of a major earthquake in our lifetime is slim, it does exist, and it would be prudent for us to be prepared, perhaps by enacting appropriate building codes and establishing emergency response procedures. Maybe we'll be lucky and nothing will happen.

4

Plate Tectonics

The preceding chapters provide a basic background to the science of geology. They describe how rocks form and how they are deformed, and they explain the fundamental principles of geologic time. Missing at this stage is an explanation for why these things happen. Why are volcanoes present at some places and not at others? Why do some places on the earth experience compression and others extension?

The driving force behind the dynamic earth—in effect, the key to understanding how the earth works—is the theory of *plate tectonics*. As a unifying theory of science it is a relative newcomer, having been developed in the 1960s and 1970s. It is beyond the scope of this book to discuss all of the details of plate tectonics. However, a few basic principles of the theory will help us in our goal of unraveling earth history.

The fundamental idea behind plate tectonics is that the earth is divided into a dozen or more rigid *lithospheric plates*, which are in motion with respect to each other (Figure 11a). The plates are about sixty miles thick, small by comparison to the eight-thousand-mile diameter of the earth. Although a wide range of sizes exist, most of the plates are large enough to raft entire continents across the earth's surface.

Plates move at rates of a few inches to a foot per year. Although virtually imperceptible to mere mortals, when these rates are multiplied by geologic time they can result in substantial displacements. For example, at a rate of about six inches per year a plate could move from the equator to the North Pole in only sixty million years, a short time geologically. The engine driving plate motion is thought to be convection cells in the mushy layer immediately beneath the plate called the *asthenosphere*. A convection cell is created when hot material moves upward, because it is less dense than the surrounding material, while cooler, denser material sinks. It is the interaction between the moving plates that is ultimately responsible for geologic phenomena, such a volcanism, earthquakes, and mountain building.

Plates interact in three ways: (1) by moving away from each other, called *divergence*; (2) by moving toward each other, called *convergence*; and (3) by moving parallel to each other. (Figure 11b). When plates diverge, a new plate is created in between them in the form of an ocean basin. As the plates pull apart, the underlying asthenosphere rises upward and, as a result of the decrease in pressure, begins to partially melt. The basaltic magma created by the partial melting rises to the surface, some of it erupting on the sea floor as basalt volcanoes and the rest crystallizing below the surface as gabbro. In this way new material is added to the margins of the diverging plates. If we drained the ocean basins we would see the surface expression of this process, a line of volcanoes, called a *spreading ridge*, that marks the boundary of diverging plates.

When plates come together, or converge, one plate is forced beneath the other, a process known as *subduction* (Figure 11b). As it descends into the hot earth the subducting plate partially melts. Rising toward the surface, the magma erupts as a string of volcanoes on the overriding plate, referred to as a *volcanic arc*, because the chain of volcanoes has an arcuate or curved shape on a map. Arc volcanism produces mostly andesite and rhyolite, with diorite and granite hidden in the deep-seated roots of the volcanoes. The Aleutian Islands of Alaska and the islands of Japan are volcanic arcs, as are the continental margin volcanoes of the Cascade Range of northern California, Oregon, and Washington, and the Andes Mountains. Subduction also results in some of the world's largest earthquakes, such as the devastating 1994 Kobi earthquake in Japan.

Much less common than plate divergence or convergence is parallel motion of plates. The best example is the San Andreas fault system, where the North American and Pacific plates grind past each other. Because there is no thermal disturbance of the rocks beneath the plate boundary, there is no volca-

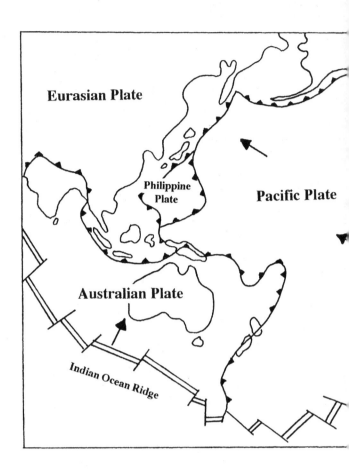

Figure 11.
A. Location and direction of motion of plates on the earth's surface. Arrows show plate motion; double lines are spreading ridges; lines with sawteeth are subduction zones.

B. Schematic cross sections of plate interactions.

Converging Plates

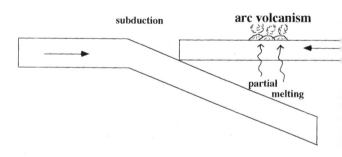

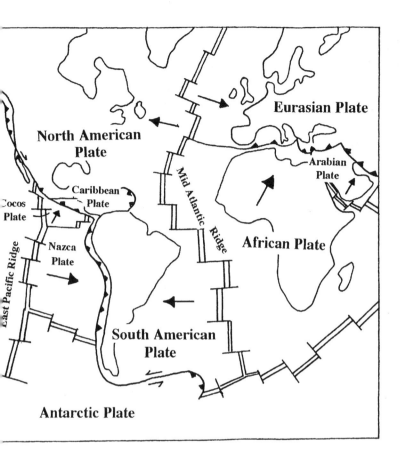

A

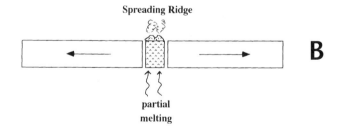

Diverging Plates

B

nism associated with parallel motion of plates. There is, however, earthquake activity, attested to by the nightmares of Californians.

Plate tectonic theory successfully explains the majority of geologic phenomena. It accounts for the major belts of volcanoes and earthquakes, as well as providing a viable mechanism to explain the presence of tropical coal beds in Illinois and polar glacial deposits in Brazil by the movement of continents across lines of latitude. Plate tectonics also explains in a less obvious way other important geologic facts. Because continental crust is too buoyant to be subducted, the oldest rocks on the earth survive on the continents. Thus, the core of all of the continents of the earth is composed of granitic and metamorphic rocks several billion years old. That is not to say that continents are immune to the disruptive effects of plate interactions. Continental crust can be ripped apart, as is currently happening in the East African rift, and continents can crash together, such as when India recently collided with Asia, creating the Himalaya Mountains. In contrast to continents, it is the destiny of oceanic crust to be consumed at subduction zones and re-created at spreading ridges. Consequently, rocks that make up the ocean basins are no older than Jurassic, a mere 150 million years or so. Plate tectonics is also ultimately responsible for the rise and fall of sea level on a global scale. Plate motion carries continents over the poles, allowing the formation of glaciers, whose growth and melting causes sea level to rise and fall thousands of miles away. Periods of active plate motion and consequent growth of volcanoes at spreading ridges can also cause sea level to rise by a process analogous to placing a large boulder in a bathtub. During periods when plate motion is rapid and new oceanic crust is created quickly, volcanoes at the spreading ridges are large, displacing water from the ocean basins onto the continents. Thus, even areas far from plate boundaries experience the effects of plate tectonics as they are inundated by the sea or as it slowly drains away.

Lest we assume plate tectonics fully explains the geologic world, there are a few geologic phenomena that do not conform well to the model. Plate tectonics predicts that deformation and earthquakes should occur at plate boundaries, yet there have been large earthquakes in the interior of plates, such as the devastating earthquakes of 1811–12 near New Madrid, Missouri. Another feature not anticipated by plate tectonic theory is *hot spots*, which are places far from plate boundaries where magma is generated for millions of years at a time. The Hawaiian Islands are the result of a hot spot beneath the center of the Pacific plate, and Yellowstone Park is situated above a continental version of a hot spot. Although they are difficult to explain in the context of plate tectonics, hot spots provide an unexpected benefit to the model. As plates move over hot spots a chain of volcanoes is created that tracks the absolute direction and rate of plate motion for millions of years.

One final thought on plate tectonics. The plates that exist today are not necessarily the ones that existed in the geologic past. As plate configurations change through time, any given region of the earth can be subjected to a variety of different geologic processes. So don't be surprised to see arc volcanic rocks, tropical limestones, and the remnants of collisional mountain ranges in the rock record of west Texas and southern New Mexico.

PART II

A Brief History of West Texas and Southern New Mexico Before Man Arrived

The goal of part II of this book is to present in simple form the geologic history of west Texas and southern New Mexico. This history, of course, is based on examining and understanding the rock record. Before beginning it is necessary to define two new terms, geologic formation and paleogeographic map.

Geologists divide the sequence of rocks in any given area into easily recognizable units called *formations*. Designating formations makes it much easier to discuss the geology, as well as providing the basic unit of geologic maps. The only rules about formations are that they must be easily distinguishable from the other formations in the sequence; they must have a known "relative" position in the sequence of rocks; and they must be thick enough, generally thicker than one hundred feet, to be mapped on standard topographic maps. A formation need not be composed of a single type of rock, although some are; nor is it necessary for the absolute age of the formation to be known at the time it is designated. A formation is named for the geographic location where it is best exposed or was first described, and the names have only local usage. For example, formation names in west Texas and southern New Mexico are not the same as those in northern New Mexico, central Texas, or southeastern Arizona. The first geologists to work in an area have the honor of naming the formations, which was accomplished in our region within about twenty years on either side of the turn of the twentieth century. There are several dozen formations in west Texas and southern New Mexico, and we will discuss most of them in the forthcoming chapters and in the context of the field trips.

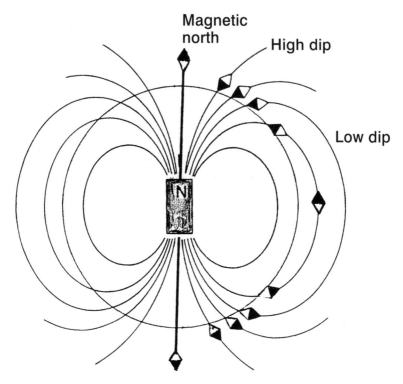

Figure 12. The magnetic force lines of the earth. The inclination is the angle at which a compass would point into the ground or up into the air, if it were allowed to swing freely about a horizontal axis. The inclination depends upon the latitudinal position of the compass.

A convenient way to illustrate geologic history is with *paleogeographic maps*. These maps are an interpretation of the landscape of a region during a particular interval of geologic time. They show the location of mountains, the position of the shoreline, the flow of rivers, or anything else that is worthy of consideration. A paleogeographic map is a "best guess" of the conditions that existed at a certain time, arrived at by interpreting the origin of the available rocks of the appropriate age. There is never complete geographic coverage of a region by outcrops or subsurface wells, however, and the gaps in the data must be filled in by extrapolation and intuition. It is much like putting together a puzzle without all of the pieces. As more data become available or interpretations of the rocks change, the paleogeographic maps change as well.

One important aspect of a paleogeographic map is interpretation of the paleolatitude, that is, where the region of interest was in a specific time interval with respect to the equator and the poles. This interpretation is especially critical to understanding earth history, because climatic zones are largely a function of latitude. For example, the climate is warm and wet near the equator, warm and dry at around 30 degrees north and south latitude, moderately moist with warm summers and cold winters in the midlatitudes, and cold and dry near the poles. Climate naturally affects the kind of sedimentary rocks that form, but is also a key factor in determining the types of organisms that live in a particular zone. Fortunately, there is a way to determine fairly accurately the paleolatitude at which ancient rocks formed by the phenomenon of *paleomagnetism*. Iron atoms in rocks act like tiny magnets. At the time the rock forms, the iron atoms align in the direction of the earth's magnetic field. They not only point in the direction of the magnetic pole, which, by the way, has flip-flopped from north to south many times through earth history, but they also point down into the ground or up in the air at an angle parallel to the lines of magnetic force (Figure 12). It is this angle, or inclination, that is the key to deciphering the paleolatitude. As you will see in the paleogeographic maps that follow, our area has changed latitude dramatically through the last billion years.

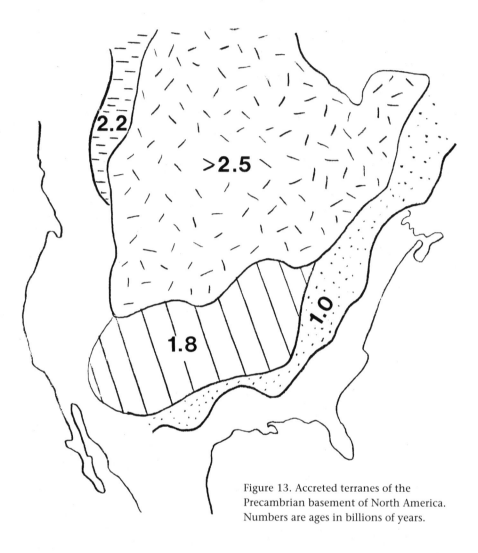

Figure 13. Accreted terranes of the Precambrian basement of North America. Numbers are ages in billions of years.

5

Precambrian: Collision and Accretion

The Precambrian history of North America and other continents of the world is primarily one of *continental accretion*, which is the successive addition of crustal material to the edges of a continent by means of collisional events. Continental accretion is analogous to a child building an object out of clay. Starting with a single lump, the object grows larger by slapping additional clay pieces onto the edges until the desired shape is attained, usually an ash tray, following the final touch of pressing a thumb into the center. So, too, North America has been slapped together by continental accretion and awaits a giant meteor collision near Minneapolis to become a giant ash tray.

The process of continental accretion begins with sediment, and perhaps volcanic rocks, deposited along the continental margin. Eventually, exotic terranes of rock, perhaps volcanic arcs, oceanic plateaus, slivers of continental crust, or even other continents, crash into the edge of the continent. The previously deposited sedimentary and volcanic rocks, as well as rocks of the colliding element, are metamorphosed and deformed into folds and thrust faults as they become part of the new continental margin. A geologically young example of this process is the island of New Guinea, which has recently been added to the northern margin of the continent of Australia by means of a volcanic arc-to-continent collision. In many cases the accreted crust is thick enough that the bottom of the pile begins to partially melt, generating magma that rises upward and crystallizes as granite. It is quite common, then, for collisional terranes composed of metamorphic rocks to be cross cut by granite only slightly younger than the metamorphism.

The accretionary history of North America is clearly evident in the map of Figure 13, which shows the age in billions of years, determined by radioactive decay, of the Precambrian basement rocks. The term *basement* refers to the metamorphic and granitic foundation of the continent, which underlies the largely unmetamorphosed sedimentary and volcanic "cover" rocks of Phanerozoic age. The original core of North America was in central Canada, with accreted rocks becoming progressively younger outward. Note that the region of west Texas and southern New Mexico was not accreted to North America until about one billion years ago, well into the Proterozoic Era. We are fortunate here, because younger deformational events have uplifted and exposed Precambrian basement rocks for us to see. When I was a geology student in Indiana, it was well known that Precambrian basement was present only a mile from campus—unfortunately it was a mile straight down!

As you examine Precambrian metamorphic rocks in the field it is interesting to speculate on their complicated history. What were the rocks before they were metamorphosed? Under what conditions of temperature and pressure were they metamorphosed? What type of collision was responsible for their accretion? These are the same questions asked by the geologists who spend their careers studying these rocks. The answers are often elusive.

Finally, what type of life existed in the Precambrian? From about 3.5 billion years to near the end of the eon life was restricted to microscopic bacteria and cyanobacteria. The latter organism, also called blue-green algae, played an important role in earth history by supplying oxygen to the atmosphere through the act of photosynthesis. We breath oxygen from the air, because cyanobacteria were around long enough to put it there. Throughout most of Precambrian time, the seas were devoid of large plants and animals and the land was almost totally barren of life. It was not until the end of the Proterozoic Era that the first multicellular life evolved in the sea, although these organisms left a poor fossil record because they had no hard parts. The only fossils in the Precambrian rocks of west Texas and southern New Mexico that I am aware of are cabbage head–shaped mounds, called *stromatolites*, representing the colonies of cyanobacteria. With the dawn of the Paleozoic Era, however, the fossil record was to improve greatly.

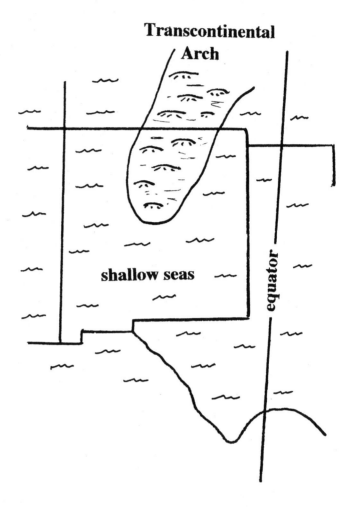

Figure 14. Early and middle Paleozoic (Cambrian through Mississippian Periods) paleogeography of New Mexico and west Texas. Note that North America was near the equator and oriented 90 degrees to its present position (turn the page 90 degrees clockwise to orient the equator in its proper position).

6

Early and Middle Paleozoic: Shallow, Subtropical Seas

Following a tremendous gap (unconformity) of almost 550 million years, the rock record in west Texas and southern New Mexico continues with sedimentary rocks of early and middle Paleozoic age (Cambrian through Mississippian Periods). During this time period, which lasted almost 200 million years, several conditions combined to favor deposition of shallow marine sediment, mostly limestones. The region of west Texas and southern New Mexico at this time was positioned far from plate boundaries and was tectonically quiet. Thus, there was no volcanism or mountain building in the region. However, active sea floor spreading and complementary growth of oceanic volcanoes in other parts of the world had the effect of periodically causing sea level to rise onto the stable parts of the continents of the world, including our own. In west Texas and southern New Mexico these seas were warm, tropical seas, because at that time our region was situated very near the equator (Figure 14). These warm seas nourished abundant life, whose shells accumulated to make limestone (Figure 15). The intervals of deposition of shallow marine sediment were interrupted by periods of erosion, initiated when worldwide sea level dropped and the shallow seas drained away from the interior of the continents. The net effect of the rise and fall of tropical seas is a stratigraphic record of shallow marine sedimentary rocks separated by unconformities (Figure 16). The only Early and Middle Paleozoic formations not composed primarily of limestone and dolomite are the sandstone-rich Bliss Formation of Cambrian age and the mixed shale and limestone-bearing Percha Formation of Devonian age (Figure 16).

Figure 15. Common Paleozoic invertebrate fossils. Bar scale represents one inch.

PERIODS	FORMATIONS	
MISSISSIPPIAN	Helms / Rancheria / Lake Valley	ls
DEVONIAN	Percha	sh, ls
SILURIAN	Fusselman	dol
ORDOVICIAN	Montoya	ls, dol
	El Paso	ls, dol
CAMBRIAN	Bliss	ss, sh, ls

Figure 16. Early and middle Paleozoic stratigraphy of west Texas and southern New Mexico.

ls=limestone;
sh=shale;
dol=dolomite;
ss=sandstone.

of marine invertebrates (animals without backbones) that are alive today, and even some now extinct, had appeared on the earth. It was an explosion of life unlike any in earth history. Perhaps just as important, the marine invertebrates developed the ability to make hard parts — shells composed of the mineral calcite, the same mineral found in egg shells. It is not clear why this happened when it did. Was it a response to increased competition? Did it reflect a change in ocean chemistry? These basic questions are difficult to answer, but make for enjoyable speculation.

As a result of the "explosion of life" and the acquisition of shells in the Cambrian Period, the early and middle Paleozoic shallow marine limestones and dolomites of our region contain abundant fossils. Particularly fossiliferous are the El Paso and Lake Valley Formations, which are featured in several of the field trips. The most common fossils are brachiopods, corals, crinoids, and bryozoa (Figure 15). Less common, but present nonetheless, are cephalopods, trilobites, and sponges.

Figure 17. Late Paleozoic (Pennsylvanian and Permian Periods) paleogeography of the Ancestral Rocky Mountains. In upper diagram, mountain ranges are shown in black.

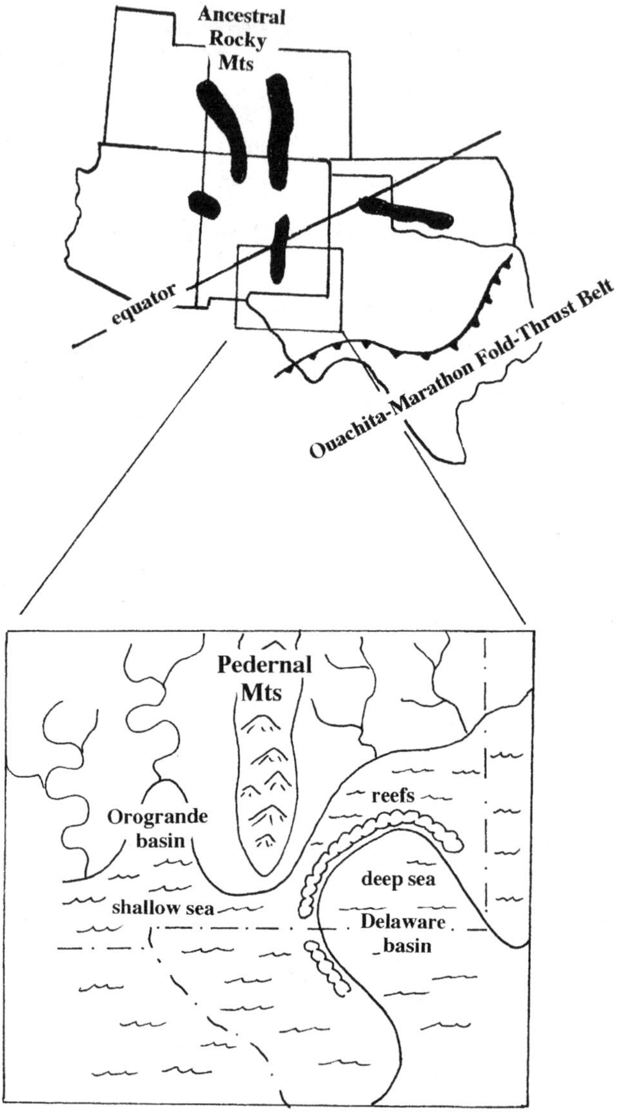

7

Late Paleozoic:
The Ancestral Rocky Mountains

The quiet conditions of the early and middle Paleozoic were broken in late Paleozoic time (Pennsylvanian and Permian Periods) by a major mountain-building event called the Ancestral Rocky Mountains, which not only affected large parts of Texas and New Mexico, but Colorado and Utah as well (Figure 17). This deformational event had a profound effect on the rock record of our region, resulting in Permo-Pennsylvanian sedimentary rocks as thick as all of the earlier Paleozoic rocks combined, and producing the first nonmarine rocks in the stratigraphic record.

Three tectonic features, an uplift and two complementary basins, played an especially important role in the geologic history of west Texas and southern New Mexico. The uplift was the Pedernal Mountains, whose axis was located just east of the present-day Sacramento Mountains. Cumulative vertical movement on the border faults of the Pedernal Mountains was at least a half-mile, enough to bring Precambrian basement rocks to the surface. Although the exact size cannot be determined, the Pedernal Mountains probably stood as high as do mountains of west Texas and southern New Mexico today.

As the Pedernal range rose, areas adjacent to it subsided and accumulated sediment. These areas were the Orogrande and Delaware basins. A complex mosaic of sediment and depositional environments existed in the basins. Conglomerates containing boulders of Precambrian basement rocks were deposited directly adjacent to the uplift and are exposed today in the Sacramento Mountains. Large rivers also flowed southward from source areas in northern New Mexico and Colorado. These rivers, which entered the sea near Las Cruces, deposited thick beds of siltstone and sandstone in channels and shale in adjacent floodplains. These nonmarine rocks, called the

Abo Formation, are easily recognizable because of their brick red color (Figure 18).

Shallow seas, in which invertebrate life flourished, occupied substantial parts of both basins and deposited marine limestone. Especially interesting to students of geology are limestones and dolomites of the Permian reef complex, which rimmed the Delaware basin and are beautifully exposed in the Guadalupe Mountains, including Carlsbad Caverns. Marine fossils in Permo-Pennsylvanian rocks are similar to those of older Paleozoic formations, including brachiopods, bryozoa, corals, crinoids, and sponges. Late in the history of the Ancestral Rockies, as the mountains were wearing down and the basins were drying up, windblown sand and later shallow seas covered the denuded crest of the range and thick deposits of gypsum and halite precipitated within the basins. Huge rooms excavated from beds of salt thousands of feet below the surface east of Carlsbad, originally excavated in the process of potash mining, have been designated as potential repositories for nuclear waste, the famous or infamous—depending on your point of view—WIPP site (Waste Isolation Pilot Project).

Particularly fascinating to me is the complex mix of depositional environments deposited along the early Permian shoreline and featured in Field Trip 8 in the Doña Ana Mountains north of Las Cruces. At this and other locations are beds of red siltstone, tan dolomite, and shale with calcite soil nodules deposited on a broad, tide-dominated shoreline, gray shale deposited just offshore of low tide, and gray fossiliferous limestone deposited in the clear, shallow sea. These kinds of sedimentary beds, each a few feet thick, are repeated over and over for hundreds of feet of thickness, suggesting that sea level rose and fell on a time scale of hundreds of thousands of years. One explanation for the fluctuating sea level was the waxing and waning of glaciers on the continent of Gondwanaland, which included the modern-day continents of Africa, South America, India, Australia, and Antarctica, whose southern tip was positioned over the south pole. Despite being thousands of miles away, the Gondwana glaciers apparently exerted a strong influence on sediment deposited in the tropical seas of west Texas and southern New Mexico.

PERIODS	FORMATIONS	
PERMIAN	Salado gyp	Castile gyp
PERMIAN	Artesia Group ls, dol, gyp	Delaware Mts. Group ss, sh, ls
PERMIAN	San Andres	ls
PERMIAN	Yeso	ss, sh, ls, gyp, dol
PERMIAN	Abo/Laborcita ss, sh	Hueco ls
PENNSYLVANIAN	Holder ss, sh, ls, cgl	Magdalena Group ls, sh, ss, cgl
PENNSYLVANIAN	Beeman ls, sh, ss	Magdalena Group ls, sh, ss, cgl
PENNSYLVANIAN	Gobbler	Magdalena Group ls, sh, ss, cgl

Figure 18. Late Paleozoic stratigraphy of west Texas and southern New Mexico. Pennsylvanian formations in left column are present in the Sacramento Mountains, whereas the term "Magdalena Group" is used farther west. Artesia and Delaware Mts. Groups are for the Guadalupe Mts.

gyp=gypsum; cgl=conglomerate; all other abbreviations as in Figure 16.

The origin of the Ancestral Rocky Mountains is not well understood and, like so many events in geology, has been the subject of many theories. Most plausible to me is that the Ancestral Rocky Mountains were a direct result of collision between the southern margin of North America and the northern margin of Gondwanaland. Despite being centered hundreds of miles to the south, the collision is thought to have transmitted stress well into New Mexico, Colorado, and Utah, just as the Himalayan collision has caused deformation well into Asia.

Another important consequence of the continental collision mentioned above is creation of the supercontinent of Pangaea, which included all of the continents of the world united. This huge continent had a significant influence on global climate, not only in the form of glaciers at its southern tip, but also in making the interior of the continent drier than normal. A dry climate is evident in floodplain deposits of the Permian Abo and Laborcita Formations by the presence of *caliche,* a form of soil rich in calcite that only exists in relatively arid regions of the earth. The dry climate also profoundly affected the plant and animal life on land. Not evolving until late Silurian time, early land plants were dominated by spore-bearing varieties, such as ferns, which require a wet environment to reproduce. As the climate dried in the Permian, dominance passed to primitive conifers, which had seeds that were not dependent upon standing water to reproduce. These early conifers can be found, although sparingly, as fossils in the Abo Formation. The climate change also benefited those land-dwelling animals who could survive in a more arid landscape. Amphibians, the first animals with backbones to walk upon the earth in latest Devonian time, surrendered their dominance to reptiles and mammal-like reptiles in late Paleozoic time. As was the case with spore-bearing plants, the need of amphibians to reproduce in water was a liability in the drier climate, and rendered them less competitive than those creatures who had developed the protected egg. Tracks of these vertebrates have been found in the Abo Formation at a variety of spots and by a variety of people in New Mexico.

8

Early Cretaceous: The Chihuahua Trough

Rocks of Triassic and Jurassic age are not common or well exposed in west Texas and southern New Mexico, although some strata of Jurassic age are present in mountain ranges along the Rio Grande northwest of Big Bend and may be present in ranges within the boot heel of southwestern New Mexico. There are, however, excellent exposures of Triassic and Jurassic rocks north of our region in the Colorado Plateau, centered around the Four Corners, where New Mexico, Arizona, Utah, and Colorado share a common point. These rocks include river sediment and volcanic ash of the Chinle Formation, beautifully exposed at Painted Desert and Petrified Forest National Parks, and crossbedded windblown sandstone of the Navajo and Entrada Formations exposed in Arches, Canyonlands, and Zion National Parks. Also of interest are dinosaur-bearing river deposits of the Jurassic Morrison Formation exposed at Dinosaur National Monument in northeastern Utah, and elsewhere in the Colorado Plateau.

The Mesozoic rock record in west Texas and southern New Mexico begins in earnest with Early Cretaceous sedimentary rocks deposited in a basin referred to as the Chihuahua trough (Figure 19). From its beginning in the Gulf of Mexico, the Chihuahua trough extended northwestward across the southernmost part of New Mexico into southeastern Arizona, where it changes name to the Bisbee basin. The Chihuahua trough/Bisbee basin is thought to be the result of extensional stress, either associated with the creation of the Gulf of Mexico or as a result of stress developed behind a volcanic arc in Mexico. Lower Cretaceous rocks are well exposed, but unfortunately difficult to access, in Sierra Juarez, in the East Potrillo Mountains thirty miles west of El Paso, and in the Big Hatchet, Little Hatchet, and Peloncillo Mountains

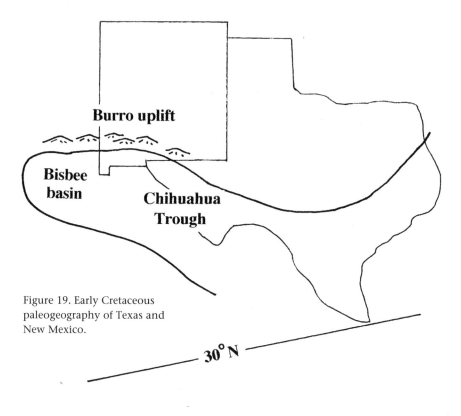

Figure 19. Early Cretaceous paleogeography of Texas and New Mexico.

of southwestern New Mexico. A much more convenient place to see these rocks is along the flanks of Cerro de Cristo Rey (Field Trip 11).

A tremendous amount of Lower Cretaceous sediment was deposited in the Chihuahua trough, with up to twelve thousand feet of strata exposed in the Big Hatchet Mountains. Represented are both shallow marine and nonmarine environments, including a wide range of sedimentary rock types: conglomerates, sandstones, siltstones, shales, and limestones (Figure 20). The sources of detrital sediment were fault-block mountains scattered throughout the basin and a major range called the Burro Uplift located north of the basin, where Paleozoic and locally Precambrian rocks were uplifted and eroded. Marine deposition occurred in response to periodic incursion of Gulf waters along the axis of the basin.

The terrestrial fossil record of Early Cretaceous age in our region is sparse and restricted to poorly studied petrified wood. Consequently, there is little hard

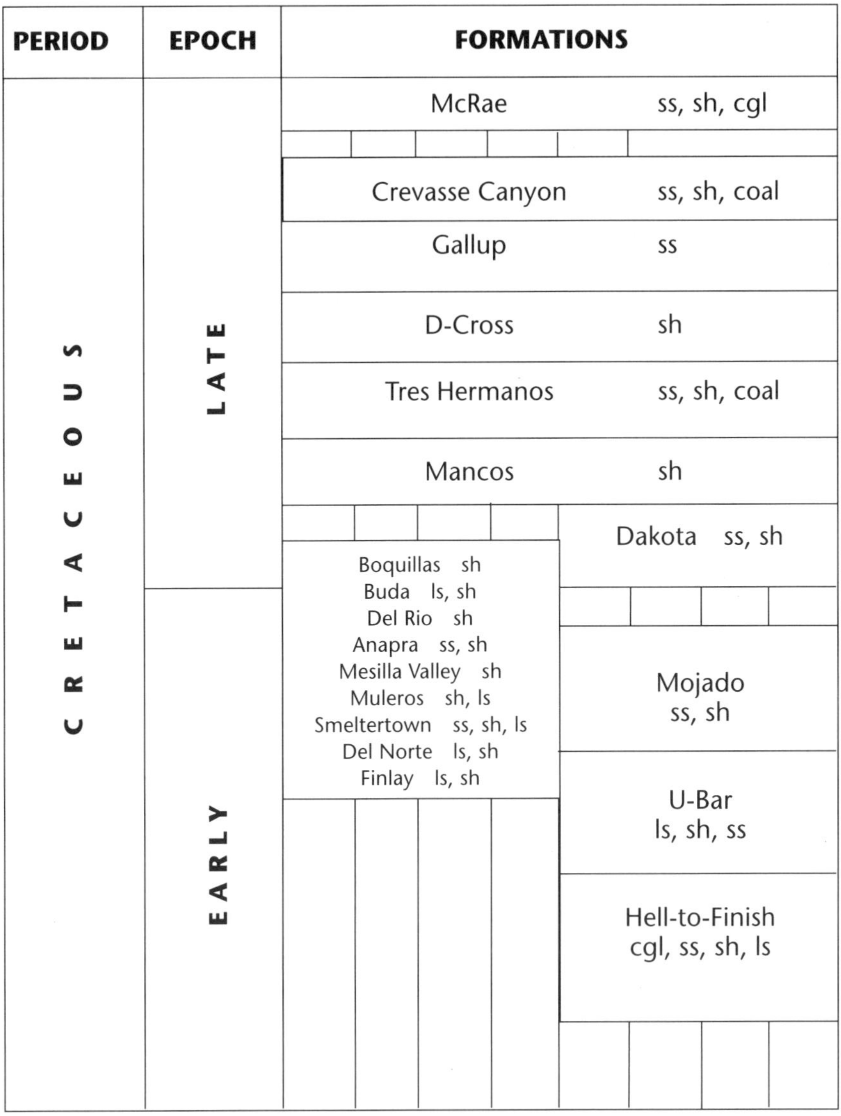

PERIOD	EPOCH	FORMATIONS	
CRETACEOUS	LATE	McRae	ss, sh, cgl
		Crevasse Canyon	ss, sh, coal
		Gallup	ss
		D-Cross	sh
		Tres Hermanos	ss, sh, coal
		Mancos	sh
	EARLY	Boquillas sh / Buda ls, sh / Del Rio sh / Anapra ss, sh / Mesilla Valley sh / Muleros sh, ls / Smeltertown ss, sh, ls / Del Norte ls, sh / Finlay ls, sh	Dakota ss, sh / Mojado ss, sh / U-Bar ls, sh, ss / Hell-to-Finish cgl, ss, sh, ls

Figure 20. Cretaceous stratigraphy of west Texas and southern New Mexico. Finlay through Boquillas Formations are used in west Texas, including Cerro de Cristo Rey; Hell-to-Finish through Mojado Formations are used in southwestern New Mexico. Dakota through McRae Formations are found in the Caballo Mountains. Abbreviations as in Figures 16 and 18.

information about the fauna and flora that inhabited the west Texas and southern New Mexico landscape, although Early Cretaceous fossils from surrounding areas suggest that dinosaurs roamed through open forests of conifers, ferns, and cycads (a tree similar in appearance but not related to palms). In contrast, the marine invertebrate fossil record is excellent, providing a clear picture of the organisms inhabiting the shallow subtropical seas of the Chihuahua trough.

The marine invertebrates of the Mesozoic seas, including the Early Cretaceous Chihuahua trough, directly benefited from a pulse of mass extinction occurring at the boundary between the Permian and Triassic Periods. This mass extinction was the greatest in earth history, far surpassing the extinction at the end of the Cretaceous Period in numbers of species lost. Gone were many common marine invertebrates, such as trilobites and horn corals. Even those that survived, such as brachiopods, crinoids, and bryozoa, would never again reach the dominance they enjoyed in the Paleozoic. The main benefiters of the mass extinction were members of the Mollusks: clams, oysters, snails, and cephalopods, the latter including squidlike animals, the ammonites, who lived in ornate, coiled shells. One type of clam that was common in but would not survive the Cretaceous was rudistids, which mimicked the shape of the extinct horn corals and built massive reefs. The Mollusks dominated Mesozoic seas, just as they do modern ones. It is because of the Permian mass extinction that airport lounges have oyster bars and not brachiopod bars.

9

Late Cretaceous:
The Western Interior Seaway

Throughout most of Mesozoic time the western margin of North America experienced a plate-tectonic setting much like that of western South America today. A small oceanic plate, the Farallon plate, was moving eastward and subducting beneath North America, just as the oceanic Nazca plate currently subducts beneath South America. Diorites and granites, the plutonic roots of the volcanic arc generated by partial melting of the Farallon plate, are exposed today in the Sierra Nevada of California, including Yosemite National Park (Figure 21). By Late Cretaceous time, and perhaps some time before, compressional mountain building began east of the arc, resulting in folding and thrust faulting of sedimentary rocks in a sinuous belt running from British Columbia to southern Nevada, where it may have changed to a southeasterly trend across Arizona and into Mexico. The structures created by this deformation are spectacularly displayed in Banff and Jasper National Parks of Canada, and at Glacier National Park, Montana.

East of the compressional mountain range was a rapidly subsiding basin, referred to as a foreland basin, in which was deposited detritus eroded from the adjacent mountains. As it subsided, the foreland basin was invaded by narrow arms of both the Arctic Ocean and Gulf of Mexico. Several times during Late Cretaceous time the oceanic arms met, creating a continuous seaway, called the Western Interior Seaway (Figure 21). During these intervals Cretaceous fish could swim unimpeded from the Arctic Ocean to the Gulf of Mexico, which they probably did—the original "snowbirds." The Western Interior Seaway marked the very last invasion of the sea into the interior of North America.

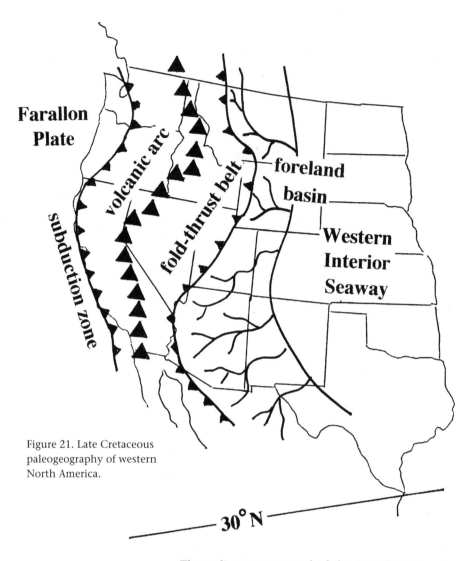

Figure 21. Late Cretaceous paleogeography of western North America.

The sedimentary record of the Late Cretaceous foreland basin in west Texas and southern New Mexico is largely restricted to sandstone, siltstone, and shale, but includes a variety of different depositional environments (Figure 20). Sinuous rivers carried sediment from the mountains toward the seaway, depositing some of it along the way. Upon reaching the sea, sand was deposited as deltas and on beaches, very much like what is happening on the Gulf Coast of Texas today. Farther out to sea, silt and clay settled out of the water, forming layers of siltstone and organic-rich black shale. There is even evidence in these shales of explosive volcanism in the arc far to the west. Yellowish layers a few inches thick of volcanic ash can be found in the marine shale. Called bentonites, because the ash has been altered to clay, they can be dated by the process of radioactive decay and provide important evidence about the "absolute" age of the rocks in which they are encased.

The theme of sea-level rise and fall introduced in the Paleozoic applies equally well to the Western Interior Seaway. The position of the shoreline in Late Cretaceous time moved back and forth inches per year. In New Mexico, the shoreline, which was generally oriented in a northwest-southeast direction, moved diagonally back and forth across the state several times. As a consequence of rising and falling sea level, Upper Cretaceous strata consist of vertical alternations of river, shoreline, and shallow marine sediment (see Field Trip 12). At least a half-dozen cycles of nonmarine to marine and back to nonmarine sediment are present in the three thousand or so feet of Upper Cretaceous rocks exposed in New Mexico. Because there is no evidence anywhere on the earth for Cretaceous continental glaciers, fluctuations in sea level must have resulted from variable rates of plate growth at spreading ridges.

Sedimentary rocks deposited in and around the Western Interior Seaway provide more than just fun and an interesting challenge to geologists like me. They also are important to the economy of New Mexico and other Rocky Mountains states. Coastal swamps supplied abundant plant debris, which, when buried, turned into coal deposits mined in northwestern New Mexico. Burial and alteration of organic matter in the dark marine shales also created oil and natural gas, which migrated out of the shale and into permeable shoreline sands, forming the important hydrocarbon reserves of the San Juan basin. The extraction of coal, oil, and natural gas not only provides jobs in northwestern New Mexico, but contributes substantial tax dollars to the state coffers.

The fossil record provides good evidence about the conditions existing in west Texas and southern New Mexico in Late Cretaceous time. The Western Interior Seaway was populated primarily by clams, oysters, snails, and a type of cephalopod called ammonites that had flowery patterns on their shells, for reasons discussed in the previous chapter. Ranging from a few inches to several feet in diameter, ammonite shells supply the most accurate information about the relative age of the rocks in which they are found. Largely absent from Late Cretaceous marine rocks are fossils of bottom-dwelling organisms, like corals, that only survive in normal marine waters. Perhaps there was too little oxygen available on the sea floor due to poor circulation, or the salinity was abnormal, or the sea floor was too soft for these organisms. Once again the cause is elusive.

The kings of the land in Late Cretaceous time, of course, were the dinosaurs. Late Cretaceous dinosaur fossils have been collected in northern New Mexico and from the Big Bend region of Texas, but are rare in our region, restricted to a few poorly preserved specimens collected from private land—and because it's private land, I can't tell you where it is. Some of these fossils, including tyrannosaurid teeth, a ceratopsian horn, and other miscellaneous bones, are on display at the Kent Hall Museum on the campus of New Mexico State University.

The most important evolutionary advance in the Cretaceous was the appearance of angiosperms, the flowering plants. These ancestors of the sycamore and willow first appeared in the mid Cretaceous, and within a few million years had become the dominant form of terrestrial plant life. Plant fossils from southern New Mexico, some of which are also on display at Kent Hall Museum, indicate our region was covered with an open-canopy forest of conifers, including relatives of sequoias, as well as cycads, palms, and a variety of broadleafed flowering trees. This assemblage of plants suggests a moderately moist, warm climate in which the temperature rarely, if ever, fell below freezing. Although not the lush tropical rainforest commonly depicted in books on dinosaurs, the woodlands of the Late Cretaceous were a far cry from the desert of today.

10

Early Tertiary: Laramide Mountain Building and Volcanism

As the Cenozoic Era dawned, the conditions that prevailed in west Texas and southern New Mexico near the end of the Mesozoic Era changed dramatically. The seaway that once connected the Arctic Ocean and Gulf of Mexico drained away. The site of compressional mountain building and arc volcanism, hundreds of miles to the west in Cretaceous time, migrated into our region. Even those organisms that had dominated Cretaceous life did not exist at the beginning of the Tertiary Period, caught in a mass extinction at the Cretaceous-Tertiary boundary that rivaled in scale that at the end of the Permian.

Plate interactions in the first two epochs of the Tertiary Period, called the Paleocene and Eocene, were little changed from those of the Cretaceous. The Farallon plate continued to subduct beneath the western margin of North America, but at a progressively shallower and shallower angle. This seemingly trivial change had profound effects on the geology of west Texas and southern New Mexico. As a consequence of the decrease in subduction angle, arc volcanism and compressional mountain building moved into our region. What had been a subsiding foreland basin in Late Cretaceous time was converted into a landscape of volcanoes and fault-bound mountain ranges and their complementary basins (Figure 22). This phenomenon, which in some places began at the very end of the Cretaceous, not only affected our region, but also areas in northern New Mexico, Colorado, eastern Utah, Wyoming, and Montana. In those areas to the north there was little or no volcanism, perhaps because subduction was at such a low angle (5°) that magma could not be generated, but

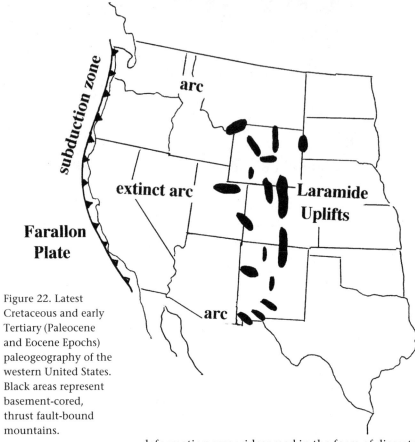

Figure 22. Latest Cretaceous and early Tertiary (Paleocene and Eocene Epochs) paleogeography of the western United States. Black areas represent basement-cored, thrust fault-bound mountains.

deformation was widespread in the form of discrete thrust fault-bound mountain ranges tens of miles wide and a hundred miles long. This deformational event is called the "Laramide," after exposures near Laramie, Wyoming. The Laramide uplifts, rejuvenated in late Tertiary time, include the Black Hills, Bighorn Mountains, Wind River Range, Uinta Mountains, and Front Range.

Laramide compressional mountain building in west Texas and southern New Mexico was similar in style to that found farther north. Thrust faults brought Paleozoic and Precambrian rocks to the surface in a series of discrete, northwest-trending ranges separated by sedimentary basins. Thousands of feet of sediment were eroded off the uplifts and deposited in the basins. All of the sediment is nonmarine, with the Western Interior Seaway having receded from the continental interior, as much because of a worldwide drop in sea level as from a local rise in the land. The Laramide sedimentary record includes conglomerates deposited at the foot of the

EPOCHS	FORMATIONS	
HOLOCENE	sand and gravel covering inset surfaces	
PLEISTOCENE / PLIOCENE	Camp Rice Palomas	ss, sh, cgl
MIOCENE	Rincon Valley	cgl, ss, sh, gyp, basalt
	Hayner Ranch	cgl, ss, sh
	Thurman	ss, sh
OLIGOCENE	Uvas	basalt

	Bell Top Kneeling Squaw Mt.
	              Nun Achenback Park
	                       Cox Ranch
	         rhyolite Cueva	

EOCENE	Palm Park Rubio Orejon
	              Peak
	           andesite	

PALEOCENE	Love Ranch Lobo	cgl, ss, sh

Figure 23. Cenozoic stratigraphy of west Texas and southern New Mexico. All abbreviations as in Figures 16 and 18.

mountains, as well as sandstone and shale deposited by rivers and in lakes (Figure 23). Laramide thrust faults and folds and coeval sedimentary rocks are exposed in a number of places in our region, but unfortunately most are difficult to access, requiring four-wheel-drive and/or being on private or government land. Only one field trip (13) encounters a Laramide fault, and it requires a difficult hike.

Unlike farther north, the Laramide event in west Texas and southern New Mexico involved volcanism. The magma was of intermediate composition, as is typical of volcanic arcs, producing gray and purplish gray andesite. Having been eroded down long ago, the volcanoes themselves are not present today, and it is not clear how many existed or where they were located. Given the thickness and widespread occurrence of their deposits, however, the volcanoes probably were similar in size to the giant stratovolcanoes of the modern Cascade Range. Although lava flows and ash flows are present within the volcanic pile, most of the deposits represent lahars, the bouldery mudflows. When you look at these rocks, think of Mount St. Helens.

The animals that roamed across the Laramide landscape, dodging lahars, were much different than those of the preceding Cretaceous Period—a mass extinction event at the Cretaceous-Tertiary boundary saw to that. Among the marine organisms to become extinct were the rudistid clams and the ammonites. The most significant casualty on land, of course, were the dinosaurs, unless you believe, as many now do, that birds represent a branch of the dinosaur tree, in which case only the non-avian dinosaurs became extinct. The untimely demise of the non-avian dinosaurs paved the way for mammals to inherit the earth. Although coexisting with dinosaurs throughout most of the Mesozoic, mammals were clearly the subordinate group, growing no larger than a possum. With dinosaurs out of the way in the early Tertiary, mammals quickly expanded into the vast variety of terrestrial and marine niches they enjoy today: from horses to whales to bats.

The cause of the Cretaceous-Tertiary extinction is controversial, which should not surprise you at this stage of the book. Considerable data has been gathered in support of the theory, first proposed in the early 1980s, that a bolide (comet or asteroid) impacting the earth at the end of the Cretaceous Period was responsible for the mass extinction event. Not everyone is convinced, however, that this impact was the sole cause of the extinctions. Doubters point to evidence for gradual extinctions of some organisms occurring over hundreds of thousands to millions of years as being incompatible with the impact theory and requiring another extinction mechanism. What that other mechanism was, however, is not clear, and the debate continues.

The best local collection of early Tertiary mammal fossils are from northwestern New Mexico. Only a few fossils have been found in Paleocene and Eocene rocks in west Texas and southern New Mexico, probably because of their inaccessibility. Plant fossils from northern New Mexico and ancient soils from our region indicate that the climate at the beginning of the Tertiary Period (Paleocene Epoch) was probably warm and wet, but became drier, although still quite warm, in the following Epoch (Eocene). The climate would continue to change, and the animals change with it, throughout the remainder of the Tertiary and into the Quaternary Epoch.

11

Middle Tertiary:
The Great Calderas

Despite lasting only about twelve million years, the middle part of the Tertiary Period, called the Oligocene Epoch, was a time of such dramatic events in west Texas and southern New Mexico that it deserves a chapter of its own. This was a time of great calderas, whose explosive products are well exposed throughout the region (Figure 23). No volcanic eruption recorded by man even comes close to those of the Oligocene.

As many as ten calderas, or collapsed volcanoes, have been identified in southern New Mexico, and several more existed in west Texas (Figure 24). Although the original shape was destroyed by later deformation and erosion, the former site of the calderas can be recognized by faults that accommodated collapse of the crater and by the tremendous volume of volcanic ash trapped within the calderas. The rocks erupted by two of these calderas, the Organ and Emory calderas, are featured in Field Trips 17 and 18, respectively. The Organ caldera is unique, in that not only are the eruptive tuffs exposed, but granites, which crystallized in the underlying magma chamber, have also been uplifted and exposed.

Rocks erupted from the Oligocene volcanoes have a bimodal composition, meaning that they represent two types at opposite ends of the compositional spectrum. The lesser mode is basalt, which erupted as lava flows. Far exceeding basalt in abundance are rhyolites, which were commonly the result of explosive eruptions. Individual eruptions created incandescent gas and ash that flowed over the countryside as ground-hugging clouds, incinerating and burying everything in their path. The white, pink, or tan-colored ash flow tuff is usually hard and dense, espe-

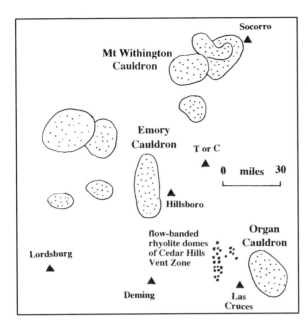

Figure 24. Location of Oligocene calderas and the Cedar Hills flow-banded rhyolite vent zone in southern New Mexico.

cially near the middle of the flow, and is characterized by flattened, fist-sized lumps of pumice. Attesting to the great magnitude of these eruptions is a fifty-foot-thick layer of ash flow volcanic tuff near Las Cruces, which can be tied to an eruption of a caldera near Lordsburg, 120 miles away! And this was not unusual. Other ash flows throughout west Texas and southern New Mexico were as far traveled and as thick. Not all of the ash escaped the caldera, though. More than ten thousand feet of ash was trapped within the Organ caldera! These eruptions were of such a magnitude that we humans can scarcely comprehend them. The blast generated by these eruptions must have blown fine ash high enough into the air to reach the jet stream, where it was distributed around the globe.

Not all of the rhyolite erupted explosively. A small amount existed as lava flows, although the lava was so viscous that it probably flowed more like cold syrup than water. Some of these rhyolite lava flows formed obsidian, a black volcanic glass. In other cases, the lava did not make it all the way to the surface, forming instead blisters on the landscape called rhyolite domes. The rocks that crystallized in these shallow domes are usually white or red and can be distinguished by fine layering called flow banding.

I do not wish to give the impression that the Oligocene was a time of continuous catastrophe. The rock record shows that there were long periods, thousands to hundreds of thousands of years, between explosive eruptions. Many successive generations of Oligocene camels never experienced a volcanic eruption. When they did occur, however, the eruptions must have created havoc unlike any experienced in modern times.

The plate-tectonic setting responsible for the great calderas is not well understood. Many believe that they represent arc volcanism in response to steepening of the angle of subduction of the Farallon plate. Others suggest that the great calderas formed in response to partial melting of the North American crust caused by the presence of the recently subducted, hot Farallon plate directly underneath it. The debate goes on.

The Oligocene fossil record in our region is sparse, as you might expect for an epoch dominated by volcanic rock. However, a few bones and trackways have been found, as well as the odd leaf imprint and even a buried petrified tree stump, now at the Geological Sciences Department at New Mexico State University. Better fossil records elsewhere in North America indicate that mammals continued to diversify, including animals almost as big as dinosaurs. The plant record suggests that a major cooling occurred at the Eocene-Oligocene boundary, a trend that would continue throughout the remainder of the Tertiary and Quaternary Periods. The architect of this cooling is thought to have been the onset of glaciation in Antarctica. We will see the effects of another glaciation a little closer to home in the Quaternary rock record.

12

Late Tertiary and Quaternary: The Rio Grande Rift

The most recent geologic event to affect west Texas and southern New Mexico is the Rio Grande rift. This interval of extensional mountain building, sedimentation, and volcanism is responsible for the landscape we see today. Extending from central Colorado to El Paso, the Rio Grande rift is a narrow gash in the earth's crust that is inexorably pulling apart (Figure 25). Rifting is far from finished, as the odd earthquake and lava flows only a few thousand years old suggest. Because of excellent outcrops, the Rio Grande rift is a natural laboratory for the study of crustal extension as well as a mecca for geologists from all over the world.

The first stage of the Rio Grande rift occurred late in the Oligocene and throughout the Miocene Epochs (Figure 23). As the crust began to pull apart, faults developed, pushing mountain ranges upward and depressing adjacent sedimentary basins. Initially, there were only a few ranges separated by broad basins floored with lakes. Owing to a dry climate, the lakes were probably ephemeral, having standing water only during the wet season. Some of these lakes were hypersaline and precipitated thumb-sized crystals of gypsum (Field Trip 20). During the beginning of rifting basaltic lava flows issued from fissures and small cinder cones, but by Miocene time there was little volcanism.

At the end of the Miocene and early in the Pliocene Epochs a second major pulse of deformation began in the southern Rio Grande rift. This pulse, which is still going on, was ultimately responsible for the present distribution of mountains and basins in our region (Figure 26). Brand new faults formed and new mountain ranges began to rise, in addition to continued activity in some of the original ranges. Many

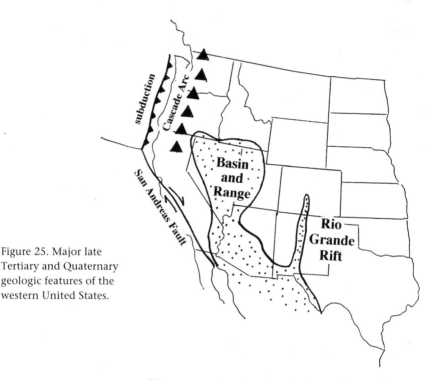

Figure 25. Major late Tertiary and Quaternary geologic features of the western United States.

of the new mountains arose in the position of former basins, uplifting and exposing previously deposited Late Oligocene and Miocene sedimentary and volcanic rocks. The result is a more closely spaced series of largely north-trending ranges and basins that dominate our modern landscape.

Sedimentation in the younger basins was similar to that in the older basins. Detritus eroded from the rising mountains was washed into the basins by flash floods. The deposits became progressively finer with increasing distance from the mountain front, changing from conglomerate to sandstone to shale. One important difference between the more recent and initial stage of sedimentation was the appearance of a river that connected many of the basins. This river, the ancestral Rio Grande, first entered our region about four million years ago and probably looked similar, although perhaps a little larger, to undammed parts of the modern Rio Grande in New Mexico (see Field Trips 20 and 22). Unlike today, however, the ancestral Rio Grande did not flow all the way to the Gulf of Mexico, emptying instead into a shallow lake near El Paso.

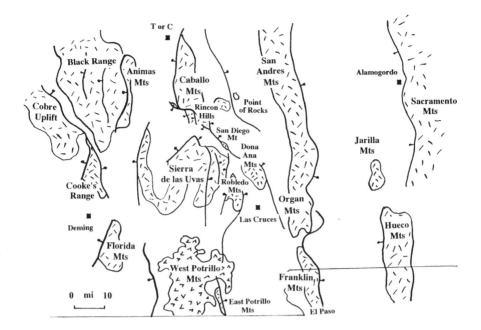

Figure 26. Location of major fault blocks of the southern Rio Grande rift. Heavy lines with ball and stick represent normal faults, with ball on the downthrown side.

Volcanism, largely absent during the Miocene Epoch, became much more common during the later pulse of rifting. In our region the volcanoes were restricted to basalt lava flows, with the greatest outpouring covering about four hundred square miles in the West Potrillo Mountains, located forty miles west-northwest of El Paso and thirty miles southwest of Las Cruces. A geologically young basalt flow can be viewed and walked on at Valley of Fires State Park near Carrizozo, New Mexico. Another spectacular and somewhat unusual example of volcanism is Kilbourne Hole on the west mesa south of Las Cruces. This mile-diameter crater resulted from a phreatomagmatic eruption, in which ascending lava intersected the water table, causing the water to flash to steam and to erupt explosively.

Despite its absence in the southern Rio Grande rift, volcanism at the rhyolite end of the compositional spectrum occurred in the Jemez Volcanic Field near Los Alamos, New Mexico. Several major ash flow eruptions rocked the volcanic field, the last one about 1.2 million years ago creating the *Valles* caldera. At least four times during the late Pliocene and early Pleistocene Epochs explosive eruptions in the Jemez Volcanic Field choked the ancestral Rio Grande with pumice. The river transported the pumice within a few days or weeks as far south as the New Mexico–Texas border, depositing it as two- to six-foot-thick beds of pumice conglomerate. Although most of the conglomerates consist of dime-sized pumice clasts, two of the beds have boulders of pumice the size of basketballs. It is probable, although not certain, that for at least some distance the pumice floated on the river.

The final chapter in the story of the Rio Grande rift also involves the river. Beginning about 780,000 years ago, give or take 100,000 or so, the ancestral Rio Grande and its tributaries began to erode down into their own alluvial plain. The river continued to alternate between downcutting and depositing sediment all through late Pleistocene and Holocene time, up to the present day. This change to erosion had three especially important consequences. First, the river now occupies an entrenched valley about three hundred feet lower than it did in early Pleistocene time. Second, the former early Pleistocene depositional surface was left high and dry when the river began to downcut. The main geologic process acting on these abandoned surfaces, which include the west mesa between Sunland Park and Las Cruces and the Jornada plain along Interstate 25 from the rest area to the Upham exit, was caliche soil formation. Consequently, these nearly flat surfaces are underlain by a dense, white soil of caliche up to ten feet thick (Field Trip 20). The third result of the entrenchment is that Pliocene and early Pleistocene sediment deposited just prior to onset of downcutting are beautifully exposed locally for geologists like me to study. Once again, nature has been kind.

Two questions about the downcutting of the ancestral Rio Grande may come to mind. The first is, why did it happen at all? One possibility is that headward (upstream-directed) erosion of the lower

Rio Grande, the separate river that emptied into the Gulf of Mexico, finally reached the lake that acted as the terminus of the upper Rio Grande. When this happened, the two rivers were united into one continuous river from Colorado all the way to the sea. However, the upper portion of the new Rio Grande was out of equilibrium with respect to the lower part and with respect to sea level, and was forced to downcut to more closely match the elevation of its new base level, the Gulf of Mexico. There are new data south of El Paso, however, to suggest that integration of the upper and lower Rio Grande took place about two million years ago, long before river downcutting began in southern New Mexico. If the new data are correct, then some other process, perhaps climate change, caused the initial river incision. The second question is why did periods of river downcutting alternate with deposition after 780,000 years ago? This was probably driven by changes in sediment load and river discharge produced by climatic fluctuations. The climatic changes were probably caused by glacial and interglacial periods in North America. So, although continental glaciers did not exist as far south as west Texas and southern New Mexico, the effects of the northern glaciers were felt in our region through climate change.

The plate-tectonic origin of the Rio Grande rift is the subject of debate. Late Tertiary continental extension not only affected the Rio Grande rift, but occurred throughout a vast area of Nevada, Arizona, southeastern California, and western Utah called the Basin and Range (Figure 25). Together, the Basin and Range and Rio Grande rift are one of the largest areas on earth of crustal extension. At the same time that the earth began to extend in the Basin and Range and Rio Grande rift, the plate boundary in California began to change from subduction to parallel motion along the San Andreas fault system. The geographic and temporal coincidence of the two events is so striking that most geologists believe creation of the San Andreas strike-slip fault regime is responsible for extension in the Basin and Range and Rio Grande rift, although there is no consensus on the exact mechanism.

The best fossil record of the southern Rio Grande rift is in Pliocene and Pleistocene sediment and consists mostly of vertebrate bones. Plant fossils are rare — mostly petrified wood — because the sedimentary environments were well oxidized and the plant material did not survive exposure to the air. Vertebrate bones indicate that the Plio-Pleistocene fauna was very different from that of today. At various times in these epochs our region was populated by giant tortoises, several species of horses, camels, giant ground sloths, and even mammoths, whose tusk and leg bone are on display at the Kent Hall Museum of New Mexico State University. Among the predators were sabre-toothed cats, dire wolves, and *Borophagus*, the bone-eating dog. As little as a few hundred thousand years ago, west Texas and southern New Mexico must have looked much like the plains of Africa. Almost all of these creatures became extinct near the Pleistocene-Holocene boundary about ten thousand years ago. Climate change associated with the end of the last glaciation was probably responsible for the extinctions, but some believe that man, newly arrived in North America, was the culprit.

PART III

Field Trips

Now that you have successfully navigated the first two parts of this book, you hopefully can more fully appreciate the geology exposed in west Texas and southern New Mexico. Toward this end, twenty-two field trips are outlined in the third part of this book. Each trip has a central theme corresponding to one of the major events in the geologic history of the region. Most trips also have rocks or structures not related to the central theme and referred to as secondary or tertiary themes. The trips are organized chronologically from oldest (Precambrian) to youngest (Late Tertiary and Quaternary), but need not be visited in that order. Many of the trips are close to one another and it is possible to do several in one day, including Field Trips 3 and 4, 8 and 9, 14 and 15, and 20 and 21.

All but one of the trips (Field Trip 5, Bishop's Cap) is accessible by car. Each trip is rated in terms of physical exertion as "easy," "moderately difficult," or "difficult." Trips designated as "easy" involve examination of roadcuts or hikes of a few hundred yards along roads or trails. "Moderately difficult" trips require either short hikes of less than a mile on steep trails, roads, or rocks, or longer hikes in excess of two miles roundtrip in relatively flat arroyos. Only two trips are rated as difficult, Field Trip 6 at Lake Valley, which involves a hike of about a mile roundtrip up and along a steep ridge, and Field Trip 13 at San Diego Mountain, which, if an offroad vehicle is not available, involves both a long hike and a climb of several hundred feet.

The descriptions, figures, and photos in this book are adequate to find and understand the geology of each of the field trips. However, you may wish to supplement the book with geologic maps, which can be purchased by mail from the Publication Division,

New Mexico Bureau of Mines and Mineral Resources, Socorro, NM 87801. Particularly helpful for the trips in this book are the New Mexico Highway Geologic Map, and two geologic maps by Seager and others: New Mexico Bureau of Mines and Mineral Resources Geologic Map 53 (1982) and Geologic Map 57 (1987).

It is important to remember a simple rule of geologic etiquette when in the field: treat the land with respect. Don't drive off of the designated roads, even if you do have a fancy four-wheel-drive vehicle. Don't leave garbage behind, don't deface the rocks, and leave gates the way you found them. If you remember this rule and abide by it, the field areas described below will be wonderful sites for people to visit and enjoy for years to come.

Field Trip 1:
Eastern Trans-Mountain Road, El Paso, TX

This trip includes a series of four roadcuts adjacent to the westbound lane of Trans-Mountain Road (Highway L375), which can be accessed from the west side of El Paso via exit 6 of I-10 and from the east side of El Paso from U.S. Highway 54. The trip begins on the far east side of the highway and proceeds uphill to the west. The rocks are entirely Precambrian in age, approximately one billion years old, and provide a good example of the variety of igneous and metamorphic rocks that make up the "crystalline basement" of our region.

The first outcrop is on the far east side of the highway, 0.9 miles west (uphill) of the turnoff to the Wilderness Park Museum. It is a long roadcut directly across from a picnic area. Extreme caution should be used at this outcrop, both because of the danger of falling rock and because of highway traffic. Do not climb the outcrop and avoid standing under unstable overhangs or steep surfaces. The outcrop is safest at the eastern and western ends. It is also advisable not to walk on the road shoulder, because traffic is fast and the outcrop is on a curve.

The rocks exposed at the first outcrop belong to two formations, the Castner Formation and the Red Bluff Formation. The Castner Formation consists of vertically oriented green and light gray marble, dark green to black slate, and a few thin layers of tan metaquartzite (Figure 27a). Some of the slate reveals what appear to be desiccation cracks on the bedding surfaces. The Castner Formation represents limestone, shale, and sandstone that were metamorphosed under conditions of relatively low temperature and pressure. Intruding the metamorphic rocks is white to pink granite of the Red Bluff Formation. The granite weathers an orange-brown color and has a distinctly knobby appearance in the natural outcrops throughout the Franklin Mountains (Figure 27b). The granite is composed of minerals large enough to be seen with the unaided eye, including

Main Theme: Precambrian basement

Access/Difficulty: Roadcuts accessed by car; easy walk

Figure 27. Trans-Mountain Road.
A. Nearly vertical dark slate and marble intruded by veins of light-colored granite.

(on facing page:)

B. Knobby, spheroidal weathering of the Red Bluff granite.
C. Coarsely crystalline phase of Red Bluff granite composed of quartz, feldspar, and dark hornblende.
D. Light-colored Red Bluff granite cross cutting darker slates of the Castner Formation.
E. Xenolith of slate surrounded by lighter-colored granite; xenolith was broken off by intruding magma, which crystallized around it.
F. View westward from stop 3 of high cliffs of the Lanoria Formation.
G. Outcrops of the Thunderbird rhyolite at stop 4.

gray, glassy quartz, white and pink feldspar, black, equant biotite, and black, elongate hornblende. On the far eastern end of the outcrop is a vein of exceptionally coarse granite with large quartz, feldspar, and hornblende crystals (Figure 27c). The western and central parts of the roadcut have excellent examples of light-colored granite cutting across the layering of the metamorphic rocks (Figure 27d). This relationship indicates that the granite was originally a liquid magma when it crosscut the metamorphic rocks and that the granite is younger than the metamorphic rocks. In some places pieces of dark metamorphic rock were broken off and completely surrounded by the crystallizing granite (Figure 27e). These pieces are called "xenoliths," which literally means "foreign rocks."

The second outcrop in the series is 0.9 miles west (uphill) of the first outcrop. Pull off into a large gravel

parking spot and examine the roadcut directly uphill from the turnout. I recommend examining only the extreme downhill end of the outcrop; the remainder of the outcrop is dangerous, because of steep, unstable surfaces and proximity to the highway. The outcrop consists mostly of greenish gray marble and a few dark bands of slate of the Castner Formation. Notable at this stop are crystals up to an inch in diameter of dark red to maroon garnet in the marble.

One mile west (uphill) of the garnet-bearing marble is the third outcrop, which consists of a small roadcut of light gray, sparkly metaquartzite of the Lanoria Formation. These rocks are composed almost exclusively of the mineral quartz and represent metamorphosed sandstone. Although not a particularly impressive outcrop, it is one of the few places where the Lanoria can be seen at road level. The thick, massive cliffs north of the highway also belong to the Lanoria Formation and can be reached by several trails (Figure 27f).

The final outcrop is 1.5 miles farther west (uphill). Turn left at a crossover just a few hundred yards beyond the pass and into a picnic area adjacent to the eastbound lane (Figure 27g). The red outcrops at the uphill end of the picnic area belong to the Thunderbird Formation and consist of slightly metamorphosed rhyolite composed of big crystals of lighter colored feldspar in a fine, red matrix. The big feldspars crystallized slowly out of the magma when it was below the surface, while the fine matrix crystallized quickly when the lava erupted at the earth's surface. The Thunderbird Rhyolite may represent the volcanic component of the same magma that crystallized the Red Bluff Granite below the surface.

Field Trip 2:
Cobre Uplift, near San Lorenzo, NM

This trip constitutes a series of roadcuts on New Mexico Highway 152 between San Lorenzo and Bayard. (Highway 152 was formerly called Highway 90 and may appear that way on older maps.) The rocks are positioned on the footwall of the Cobre fault block, which was uplifted in late Tertiary time along the Mimbres-Sarten normal fault. The fault block is tilted to the west, so when you travel westward along the highway you encounter progressively younger stratigraphic formations.

The first outcrop is on a side road called Acklin Hill Road, which intersects Highway 152 1.1 miles west of its junction with NM Highway 35 (Figure 28a). Turn right onto Acklin Hill Road and immediately park in a turnout on the right side of road. Walk down Acklin Hill Road and examine roadcuts on the left side of the road. Be very careful of the traffic, because the road is very narrow and there is little space between the road and the outcrops.

The first outcrops encountered on Acklin Hill Road are well-bedded brown sandstone and dolomite of the Cambrian Bliss Formation (Figure 16). Some of the beds contain greenish sand-sized grains of a clay mineral called glauconite, which precipitated on a shallow sea floor. Farther down the road, sandstones of the Bliss Formation are coarser grained and maroon in color and display numerous crossbeds. The Bliss Formation was deposited on the shoreline and in a shallow, tropical sea that covered much of North America in late Cambrian time.

Farther down the road outcrops of the Bliss are covered by colluvium, which consists of angular blocks of modern debris eroded from the steep slope. Around a bend in the road the Bliss reappears in the roadcuts. Eventually dark green rocks are present beneath the Bliss. These green rocks are Precambrian in age and represent a slightly metamorphosed basalt or gabbro. The contact between the Bliss and the Precambrian is a major unconformity representing

Main Theme: Precambrian basement

Secondary Theme: Early–Mid Paleozoic Sedimentation

Tertiary Theme: Laramide Igneous Activity and Ore Formation

Access/Difficulty: Roadcuts accessed by car and easy hike along highway

Figure 28. Cobre uplift.
A. Roadcut of Cambrian Bliss Formation and Precambrian basement on Acklin Road. View is eastward toward the Mimbres Valley.
B. View of Chino open pit copper mine.

almost 500 million years of geologic time not represented by rock at this location. Precambrian rocks are not present directly across the canyon, because of a fault that trends roughly parallel to the road. The same Precambrian rock is exposed throughout the remainder of the roadcut, so because of the traffic danger it is probably best to return to the parking spot.

After returning to your vehicle, drive west (uphill) on Highway 152. The steep cliffs on the north side of the highway are dolomite and limestone of the

Ordovician El Paso and Montoya Formations, with the Silurian Fusselman Formation holding up the crest of the high ridge. These rocks are fractured and faulted and crosscut by light and dark igneous rocks. This is not a very good place to see these formations, because they are dolomitized and deformed and because the steep outcrops are dangerously unstable.

Once the highway flattens out at the top of the hill, there are several roadcuts of black shale of the Devonian Percha Formation, which was deposited on a poorly oxygenated sea floor. The upper part of the Percha has a ribbony appearance due to flattened nodules and discontinuous beds of gray limestone. Fossils are not common in the Percha, because of the paucity of oxygen on the sea floor, but it is possible to find brachiopods. At several places, including roadcuts 1.5 and 2.6 miles west of the junction with Acklin Hill Road, the Percha Formation is crosscut by volcanic rocks composed of a few visible crystals of quartz and feldspar in a fine, white matrix. At 4.1 miles from Acklin Hill Road is a roadcut of diorite containing easily recognizable crystals of white feldspar, black biotite, and a small amount of glassy quartz. These igneous rocks are probably Late Cretaceous or early Tertiary in age and may be related to the igneous body near the Chino mine.

Approximately 3.5 miles after Acklin Hill Road junction is the beginning of roadcuts of gray limestone of Mississippian and Pennsylvanian age, some of which are crosscut by igneous rocks similar to those described in the previous paragraph. Many of the limestone beds contain fossil brachiopods, crinoids, and bryozoa, but they do not easily weather out of the rocks and are difficult to collect. The limestone was deposited in shallow, tropical seas.

The final stop of this trip is a large turnout on the opposite side of the road 7.1 miles from the Acklin Hill Road junction. This site offers a spectacular vista of the Chino open pit mine (Figure 28b). Copper mineralization is associated with diorites of Late Cretaceous and early Tertiary age that intruded into Paleozoic and Cretaceous sedimentary rocks. The brown cliffs above the pit are ash-flow tuffs of the Kneeling Nun Formation ejected from the Emory Caldera thirty-five million years ago.

Field Trip 3:
McKelligon Canyon Park, El Paso, TX

Main Theme: Early-Mid Paleozoic Sedimentation

Secondary Theme: Rio Grande Rift

Access/Difficulty: Natural outcrops accessed by car; easy hike on gravel trail and short climb on outcrop

McKelligon Canyon Park is located on the east side of the city of El Paso and can be reached by taking U.S. Highway 54 and exiting on Fred Wilson Avenue. Follow Fred Wilson Avenue to the west (toward the Franklin Mountains) until it turns into Alabama Street. Follow Alabama Street southward about one mile and then turn right onto McKelligon Canyon Road. Drive to the far end of the park, where the road makes a loop. Park at any of the gravel turnouts or covered picnic spots.

From this vantage point you are in a position to look back down the canyon to the south-southeast

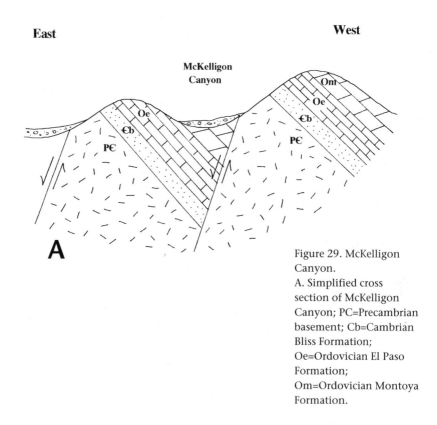

Figure 29. McKelligon Canyon.
A. Simplified cross section of McKelligon Canyon; PC=Precambrian basement; Cb=Cambrian Bliss Formation; Oe=Ordovician El Paso Formation; Om=Ordovician Montoya Formation.

B. View westward from parking spot toward the high range of the Franklin Mountains. Lower, dark, well-bedded rocks are the Bliss Formation, with weathered slopes of Precambrian granite beneath; higher, lighter-colored cliffs are the Ordovician El Paso and Montoya Formations.
C. View eastward from parking spot of Ordovician El Paso Formation dipping toward the canyon.
D. View looking south from trail of unconformity between Precambrian granite (below man's feet) and overlying Cambrian Bliss Formation.
E. View northward at same unconformity pictured in photo "D."
F. Well-bedded sandstone of the Bliss Formation.
G. Crossbeds in the Bliss Formation.

and view the stratigraphy and structure of this part of the Franklin Mountains (Figure 29a, b, c). Like most of the mountain ranges of our region, the Franklin Mountains were uplifted along normal faults during late Tertiary time as a result of crustal extension associated with the Rio Grande rift. At this location the range is separated into two blocks by a normal fault that runs down the middle of the canyon. The fault dips to the east (left, as you look down the canyon), with the upthrown side (footwall) making up the higher part of the range to the west (right, as you look down the canyon) (Figure 29a). The stratigraphy in the visible part of the footwall block includes, in ascending order, Precambrian basement, Bliss Formation, El Paso Formation, and Montoya Formation, all of which are tilted toward the west, away from the canyon (Figure 29b). The downthrown side to the east (left as you look down the canyon) consists of the same stratigraphy also tilted to the west (toward the canyon), but with the El Paso Formation exposed along the walls of the canyon (Figure 29c). Just east of this smaller block is the main border fault of the mountain range, also dipping and downthrown to the east.

The primary purpose of this trip is to examine rocks exposed in a small canyon a few hundred yards off the road. Take a gravelly trail that intersects the southwestern part of the loop road about fifty yards downslope of the last covered picnic spot. The trail goes westward into the higher part of the range. Where the gravelly part of the trail ends, there is an excellent exposure of an unconformity, with the Precambrian Red Bluff Granite below and the Cambrian Bliss Formation above (Figure 29d, e). This unconformity represents a gap in the rock record of almost 500 million years. The well-bedded, reddish brown beds of the Bliss Formation are composed of sandstone that was deposited near the shore of a tropical sea (Figure 29f). Look for feeding trails on the surfaces of the beds, as well as crossbeds produced by marine currents (Figure 29g). A modern environment similar to that in which the Bliss Formation was deposited is Padre Island on the Gulf Coast of Texas.

Field Trip 4:
Murchison Park, Scenic Drive, El Paso, TX

Murchison Park is located on the south edge of the Franklin Mountains, overlooking the cities of El Paso and Ciudad Juárez. It can be reached from the east side of El Paso by taking Alabama Street and turning west onto Richmond Avenue, which becomes Scenic Drive. Murchison Park is at the crest of the road. From the west side of El Paso, the park can be reached by following Mesa Street to Rim Road.

Two formations are exposed in the vicinity of the park, the underlying El Paso Formation and the overlying Montoya Formation, both of Ordovician age (Figures 30a and 16). Both formations were originally deposited as limestone in warm, shallow, tropical seas that covered most of North America. Modern analogs of the Ordovician seas would be the Persian Gulf and the Bahama bank. After deposition of the El Paso Formation, the seas retreated and our region experienced erosion, creating an unconformity between the two formations. Sea level rose again, however, late in Ordovician time, and the Montoya Formation was deposited.

Exposed below the parking lot, probably dropped down along a small fault, is the Montoya Formation, which consists of dark gray dolomite that replaced the original limestone after burial. Fortunately, many of the fossils and sedimentary structures survived the process of dolomitization. Especially common are burrows, which were made by marine organisms grazing along the top of or within the sediment (Figure 30b). Also present in the Montoya at this location are large fossil corals (Figure 30c), as well as brachiopods and gastropods (snails).

The El Paso Formation is best seen on either side of a trail that begins directly across from the parking lot and ascends into the mountains. Composed primarily of thinly bedded light gray limestone, the El Paso Formation has numerous burrows infilled with tan or brown dolomite or chert, a fine-grained form

Main Theme: Early–Mid Paleozoic Sedimentation

Access/Difficulty: Roadcuts and natural outcrops accessed by car; moderately difficult short climb on steep trail

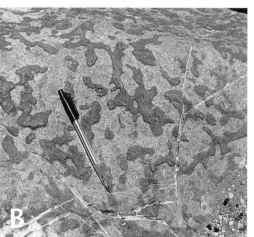

Figure 30. Murchison Park.
A. Light-colored rocks located behind truck and tilted to the left belong to the Ordovician El Paso Formation, whereas the darker cliff above is the base of the Ordovician Montoya Formation, which is also exposed below steps.
B. Dark, irregularly shaped silicified burrows in the Montoya Formation.
C. Fossil coral in the Montoya Formation.

of quartz. This formation also contains fossils, including sponges, trilobites, gastropods, and algae, but they are much more difficult to recognize, in part because many of them are broken and abraded. Also present in abundance are elongate, light-colored particles about an inch long called "intraclasts," which represent pieces of the sea floor ripped up by storms.

Field Trip 5:
Bishop's Cap, near Las Cruces, NM

This field trip can only be undertaken with an offroad vehicle. Do not attempt to visit this site by car. Take I-10 to exit 151—Mesquite and turn east (if westbound, turn right; if eastbound, turn left and cross over overpass). At the stop sign a few tens of yards from the interchange, turn right. After 0.2 miles, turn left, following the signs for the Doña Ana County Liquid and Solid Waste Disposal Facilities. Continue straight past the turnoff to the Solid Waste Disposal Facility and at Y-intersection, 2.1 miles past the last turn, veer left, following signs for the Liquid Waste Disposal Facility. Continue straight past the Liquid Waste Facility and turn right at Y-intersection, 1.7 miles after the last turn; 0.9 miles later turn right; 0.8 miles later turn right; 0.7 miles later turn right; 0.3 miles later turn right and go 0.4 miles to bottom of hill and park. Walk up road to the south toward the mine.

A westward tilt of the Paleozoic sedimentary rocks can be seen by looking southward from the parking spot (Figure 31a). The visible stratigraphy includes, in ascending order, the Silurian Fusselman Formation, the Devonian Percha Formation, the Mississippian Lake Valley, Rancheria, and Helms Formations, and the La Tuna Formation of the Pennsylvanian Magdalena Group (Figures 31a and 16). The tilt of these rocks is the result of normal faulting and block rotation during the late Tertiary Rio Grande rift. Looking in the opposite direction, to the north, it is possible to see Oligocene ash-flow tuffs that filled the Organ Mountains caldera (see Field Trip 17). The light-colored tuffs belong to the Cueva Formation and the darker rocks are part of the Tuff of Achenback Park (Figure 23).

The road toward the mine is situated on the Devonian Percha Formation, which consists of black shale and dark gray siltstone (Figure 31b). The dark color indicates the presence of finely disseminated organic matter and suggests that the shallow sea in

Main Theme: Early–Mid Paleozoic Sedimentation

Secondary Theme: Late Tertiary Rio Grande Rift

Access/Difficulty: Natural outcrops only accessed by offroad vehicle; moderately difficult hike on gravel road and on rocks

Figure 31. Bishop's Cap.
A. View looking south at the site of field trip. Rocks are tilted to the right (west); hill on left side is Silurian Fusselman Formation; road follows Devonian Percha Formation; hills to right of road are Mississippian limestones, with Pennsylvanian La Tuna Formation of the Magdalena Group holding up the ridge on the skyline.
B. Roadcut of dark gray shale and siltstone of the Devonian Percha Formation.
C. The brachiopod Lingula from the Percha Formation.
D. Limestone mound in the Mississippian Lake Valley Formation.
E. View westward from the pass of a fault that juxtaposes the Pennsylvanian La Tuna and Mississippian Helms Formations against the Mississippian Rancheria Formation.

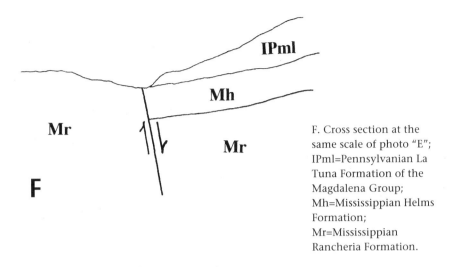

F. Cross section at the same scale of photo "E"; IPml=Pennsylvanian La Tuna Formation of the Magdalena Group; Mh=Mississippian Helms Formation; Mr=Mississippian Rancheria Formation.

which the sediment was deposited was poorly oxygenated. Stagnant conditions are also suggested by the paucity of marine invertebrate fossils. The only common fossil, best preserved in the dark gray siltstone, is the brachiopod *Lingula*, whose modern relatives live in muddy lagoons (Figure 31c). The shells of *Lingula* have a bluish appearance, because they are made of a phosphatic mineral instead of calcite.

About a third of the way up the road is a mound-shaped body of limestone in the Lake Valley Formation (Figure 31d). This type of mound, which is composed primarily of microcrystalline calcite and a few scattered crinoid plates, is common in Mississippian-age rocks throughout the world, but its origin is poorly understood. Many believe the mounds were reefs built by organisms that left no fossil record because they had no hard parts. Others suggest that the mounds originally formed by current action in shallow water and then slid into deeper water.

At the top of the hill, where the road ends, are several small mines to the east of the road. Always be careful in the vicinity of mines and never go into a mine. There is usually excellent mineral collecting in the loose piles of debris near the mine entrance. These particular mines are developed in the upper part of the Silurian Fusselman Formation, near its contact with the Percha Formation. This is a common location for mineralization, because ore-bearing fluids can easily move through the permeable dolomite of the Fusselman, but are trapped beneath the less permeable shale of the Percha. The age of the mineralization is probably late Tertiary, associated with the Rio Grande rift. Especially abundant at this site are crystals of quartz, fluorite, calcite, and barite. Quartz (silicon dioxide) is clear or gray and glassy in appearance and sometimes exists as pyramid-shaped crystals. Fluorite (calcium fluoride), which was the primary ore mineral mined here, is clear or purplish in color and displays pyramid-shaped crystals. Because they look similar, quartz and fluorite are difficult to tell apart. The best way to distinguish these two minerals is by hardness; quartz is harder than steel and cannot be scratched by a knife, whereas fluorite is softer than steel. Calcite (calcium carbonate) and barite (barium sulfate) are also difficult to distinguish, because they both exist as milky white, shiny crystals and they both have a hardness less than steel. Barite at this location tends to make more elongate crystals and it is denser, such that a sample containing barite will seem unusually heavy compared to barite-free samples of similar size. Also present is jasper, a form of microcrystalline quartz, whose red color results from the presence of small quantities of oxidized iron.

Just to the west of where the road ends is a good place to see the Mississippian Rancheria and Helms Formations (Figure 31e, f). The lower Rancheria Formation consists of thin beds of gray limestone containing dark gray blebs and bands of chert. Few fossils are present in the Rancheria, and it was probably deposited in a relatively deep sea. Climbing the western cliff, the beds of limestone become thicker and more fossiliferous, marking the change to the Helms Formation. Fossils include brachiopod shells, bryozoans, and crinoid plates (compare to Figure 15). Also present are several distinctive light brown beds composed of oolite limestone, which, along with the fossils, indicates a shallow marine depositional environment for the Helms Formation. The massive dark limestone cliffs at the top of the outcrop belong to the Pennsylvanian La Tuna Formation, and at the base of the formation are brown, silicified solitary corals.

Field Trip 6:
Lake Valley, NM

The type section of the Mississippian Lake Valley Formation, located just north of the ghost town of Lake Valley, New Mexico, is one of the best fossil-collecting sites in southern New Mexico. The natural outcrop is exposed on Apache Hill, located about 1.5 miles north of the town of Lake Valley on the east side of New Mexico Highway 27, which can be reached by going south from Hillsboro or north from Nutt. Park in a small turnout immediately south of a cattle guard 15.4 miles south of Hillsboro and 14.5 miles north of Nutt. The outcrop of interest on Apache Hill is composed of a lower slope, a prominent cliff of gray limestone, another slope above the cliff, and ledges of limestone holding up the crest of the hill (Figure 32a). The fossiliferous beds are above the cliff and below the crest of the hill. From the parking spot, the cliff is visible to the east and is accessible by climbing onto a low saddle just south of the cliff and then on to the top of the cliff. Once there, move north along the face of the hill above the cliff, but below the crest. The fossils can be found in the loose rubble on the hill slope, especially where erosion has uncovered the grayish shaley limestone (Figure 32b).

Virtually every piece of loose limestone rubble contains fossils, but what makes this site famous is the fact that individual, complete fossils weather out of the limestone and can be found among the rubble. The fossils are marine invertebrates and include several types of brachiopods, bryozoans, horn corals, and crinoid plates and heads (cf. Figure 15). Also present, but much less common, are snails, clams, cephalopods, and trilobites. In order to preserve this site for future visitors, please collect only one of each type of fossil.

There are also several small faults on Apache Hill. Crystallizing from fluids along the fault planes were large crystals of milky white calcite, which can be found among the limestone rubble.

Main Theme: Early–Mid Paleozoic Sedimentation

Access/Difficulty: Natural outcrop reached by car; difficult climb up hill

Figure 32. Apache Hill near Lake Valley.
A. View looking northeast of Apache Hill, just north of the ghost town of Lake Valley. The fossiliferous member of the Mississippian Lake Valley Formation is above the middle cliff and below the top of the ridge.
B. Weathered, rubbly slope of the fossiliferous member of the Lake Valley Formation.

Limestones of the Lake Valley Formation were deposited in a warm tropical sea that covered much of the continental United States. Throughout the western United States limestones equivalent to the Lake Valley Formation make conspicuous cliffs hundreds of feet thick. The cliff-forming Mississippian limestone makes a good marker bed to provide a "stratigraphic fix" as you drive through the West.

Field Trip 7:
Dry Canyon–La Luz Canyon, Alamogordo, NM

This trip constitutes a loop accessible by car within the Sacramento Mountains and includes five stops along the road. The primary purpose is to examine sedimentary rocks and structures formed during the Permo-Pennsylvanian mountain-building event known as the Ancestral Rocky Mountains. From U.S. Highway 54 and 70, just north of Alamogordo, turn right (east) onto U.S. Highway 82, heading toward Cloudcroft. After 3.8 miles pull off the road to the right in front of a Lincoln National Forest sign for stop 1 (Figure 33a).

Across the road from the pulloff of stop 1 and halfway up the ridge is an example of a reef developed in the Pennsylvanian Holder Formation (Figures 33a and 16). The reef was constructed on the edge of a shallow marine platform by a type of algae called phylloid algae, which became extinct at the end of the Paleozoic Era. From this vantage point the western half of the reef is visible, the portion that faced toward the open ocean. Also visible are westward-tilted flank beds, which represent debris washed off the core of the reef and deposited as an apron along the seaward side. The fossil algae are difficult to recognize in hand sample, and it is probably not worth the climb to see them in outcrop.

Continue eastward (uphill) for 2.7 miles and pull off for stop 2 onto a large gravel turnout on the opposite side of the road. The red and gray rocks in the roadcut are in the Permian Laborcita Formation (Figure 18). These sedimentary rocks were deposited by rivers that flowed from the Pedernal Mountains, one of the major mountain ranges that made up the Permo-Pennsylvanian Ancestral Rocky Mountains (Figure 17). From this location the crest of the Pedernal Mountains was ten miles or so to the east. The fine-grained red shales and siltstones were deposited on the floodplain between the river channels. Some of the redbeds have a knobby appearance,

Main Theme: Permo-Pennsylvanian Ancestral Rocky Mountains

Access/Difficulty: Roadcuts and natural outcrops reached by car; easy walks along road

Figure 33. Sacramento Mountains.
A. Phylloid algae reef in the Pennsylvanian Holder Formation at stop 1.
B. Nodules of calcite, commonly called a caliche paleosol, in red shale of the Permian Laborcita Formation at stop 2.
C. Beds of pebble conglomerate in the Permian Laborcita Formation.
D. View eastward of the Fresnal fault from stop 2; well-bedded limestones of the Pennsylvanian Gobbler Formation are on the upthrown side, whereas slopes in the foreground are on the downthrown side and are underlain by the Permian Laborcita Formation.
E. Limestones of the Pennsylvanian Gobbler Formation at stop 3.

F. Conglomerates of the Permian Abo Formation.
G. Vertical andesite dike of Tertiary age cross cutting the Permian Abo Formation.
H. Beds of the Permian Abo Formation tilted against Fresnal fault.
I. Roadcut of the Pennsylvanian Holder Formation at stop 5.
J. Limestone of the Holder Formation at stop 5 containing fusulinid fossils.

consisting of nodules of calcite that formed in a type of soil commonly called caliche (Figure 33b). This type of paleosol, or ancient soil, indicates the climate in the Permian was relatively arid. The resistant ledges are composed of conglomerate and sandstone deposited in the river channels (Figure 33c). The conglomerates contain pebbles and cobbles of older Paleozoic limestone and dolomite that were exposed in and eroded from the Pedernal Mountains, as well as clasts of caliche ripped up as the channels moved across the floodplain.

Stop 2 also provides an excellent view of the Fresnal fault (Figure 33d), which initially developed in the Pennsylvanian and Permian Periods during the deformation that created the Ancestral Rocky Mountains. The upthrown side (eastern side) of the fault brings the Pennsylvanian Gobbler Formation against the Permian Laborcita Formation on the downthrown side. It can be demonstrated that this fault was active in Permo-Pennsylvanian time, because the Holder and Laborcita Formations are absent on the upthrown side, but present on the downthrown side, and the Beeman Formation has only a fraction of the thickness on the upthrown side that it has on the downthrown side. The fault dips to the west, which makes it a normal fault. However, the fault may have been tilted eastward by late Tertiary deformation associated with the Rio Grande rift, raising the possibility that, if the rotation is removed, the Fresnal fault may have been originally a steep thrust fault.

Proceed 0.8 miles eastward to a large turnout on the opposite side of the road, immediately in front of the tunnel. From here, at stop 3, can be seen limestones of the Pennsylvanian Gobbler Formation (Figure 33e), as well as another view of the Fresnal fault. There is a trail leading to the bottom of the canyon for those who wish to get a better view of the rocks.

Travel through the tunnel and turn left on Cherry Way, a total of 1.4 miles past the turnout to stop 3. Turn right at the T-junction onto Railroad Drive. At the Y-junction near the Trinity Baptist Church go left onto a gravel road marked Mountain Park Road, and 0.3 miles farther veer left at the Y-junction onto Mountain View Road. The red rocks exposed along

Mountain Park and Mountain View Roads belong to the Permian Abo Formation and consist of pebbly sandstone and shale deposited by rivers draining the Pedernal Uplift. Crossbeds are conspicuous in many of the sandstones. Yellowish and pinkish rocks on the high cliffs above the level of the road are in the Permian Yeso Formation (Figure 18). The road descends steeply after 3.1 miles into La Luz Canyon; turn left onto La Luz Canyon Road. One mile after the turn onto La Luz Canyon Road are roadcuts of cobble conglomerates of the Abo Formation that constitute stop 4 (Figure 33f). These conglomerates contain cobbles and boulders of Precambrian metaquartzite eroded from the core of the Pedernal Mountains.

Two-tenths of a mile from stop 4 a vertical dike of andesite can be seen on the other side of the canyon (Figure 33g). Several more dikes are present along the remainder of the route. The dikes are probably early Tertiary in age. About 1.3 miles beyond the dike, beds of the Abo Formation become tilted along the Fresnal fault (Figure 33h). After another 1.3 miles, turn left at the stop sign onto Laborcita Canyon Road, and drive 2.4 miles to a pull off on the right side of the road to see exposures of stop 5 (Figure 33i). This long roadcut is in the Pennsylvanian Holder Formation and includes gray limestone and red and black shale deposited in a shallow sea. Also present are conglomerates, one bed of which is especially well exposed near the middle of the outcrop. The conglomerates occupy channels and may represent rivers or channels near the mouth of a delta. A prominent ledge-forming limestone near the eastern end of the outcrop contains fossils of tiny, single-celled, marine animals called fusulinids, which are among the best fossils for dating sedimentary rocks in the Pennsylvanian and Permian Periods (Figure 33j).

Leave stop 5, proceed 2.1 miles to a T-junction, and turn left. Go through the town of La Luz on the main road, and 2 miles farther turn right at the stop sign with a red flashing light. After 1.8 miles is the intersection with Highways 54 and 70; turn left to return to Alamogordo and right to Tularosa.

Field Trip 8:
Lucero Arroyo, Doña Ana Mountains, NM

Main Theme: Permo-Pennsylvanian Ancestral Rocky Mountains

Secondary Theme: Rio Grande Rift

Tertiary Theme: Laramide Volcanism

Access/Difficulty: Natural outcrops accessed by car and moderately difficult hike of two miles roundtrip in arroyo.

A variety of rocks and geologic features of different ages can be observed in outcrops within Lucero Arroyo. The site is nineteen miles north of Las Cruces and can be reached by taking I-25 to exit 19—Radium Springs. If northbound, turn right (east) after exiting; if southbound, turn left and cross the overpass. Park in a turnout about one hundred yards from the interchange. Walk east through the gate and head east-southeast across the desert, toward a brown cliff on the south side of the arroyo about a quarter-mile away (Figure 34a); the jagged peaks of the Doña Ana Mountains will be in the middle distance, with the larger peaks of the Organ Mountains directly behind. After several hundred yards in the desert, enter a small arroyo and turn right (south). Several hundred yards farther, when the smaller arroyo enters the main arroyo (Lucero arroyo), turn left (east) and walk up the main arroyo. All of the outcrops of interest in this trip are in or on the walls of this arroyo. The arroyo is wide, has many bends, and contains numerous vegetated "islands" of debris that may lead to confusion about the proper course. However, as long as you stay in the arroyo, it does not matter which path you take; they all eventually merge into a single channel.

The first outcrops encountered are on the north (left) side of the arroyo and belong to the Pliocene-Pleistocene Camp Rice Formation (Figures 23 and 34b). These conglomerates are composed of cobbles of gray limestone, red siltstone, and purple and white volcanic rocks eroded from the Doña Ana Mountains to the east and deposited in the northern Mesilla basin. Large flash floods were required to transport such large particles from their source. This basin was the site of deposition from about 3.5 million years ago to 780,000 years ago, at which time the ancestral Rio Grande and its tributaries, such as Lucero arroyo, began to erode into the basin, removing sediment instead of depositing it. We see this outcrop of the

Camp Rice Formation as a result of this erosion.

About a quarter-mile farther up the arroyo, exposed on cliff faces on both sides of the arroyo, is an excellent example of an angular unconformity (Figure 34c, d). Below the unconformity are gently west-tilted purple rocks of the Eocene Palm Park Formation; above the unconformity are horizontal beds of the Plio-Pleistocene Camp Rice Formation. The unconformity indicates a time gap between the two formations, during which the Palm Park Formation was uplifted, tilted, and eroded prior to deposition of the Camp Rice Formation. This unconformity represents the floor of the northern Mesilla basin, and the erosion described in the previous paragraph has cut through all of the Camp Rice Formation and into the bedrock beneath the basin. Some 780,000 years ago there would have been about three hundred feet of sediment above you at this site.

The next outcrops encountered are highly fractured tan siltstones of the Permian Hueco Formation. These rocks have been deformed against a fault. You will see better exposures farther up the arroyo.

Following a major bend in the arroyo, purple rocks are exposed on the arroyo bottom. These rocks are part of the Eocene Palm Park Formation and consist of pebbles and cobbles of volcanic andesite in a fine ash matrix (Figure 34e). They were deposited as lahars, or volcanic mudflows, that flowed down the flank of a large volcano similar to Mount St. Helens. Late in the Eocene Epoch, approximately forty million years ago, andesitic volcanism was generated in our region as a result of low-angle subduction of the oceanic Farallon Plate beneath the western edge of North America (Figure 22). Volcanic rocks of this and equivalent formations can be seen on other trips (Field Trips 14, 16, 22).

A hundred yards beyond the Palm Park outcrops are near vertical beds of the Hueco Formation, upturned against a fault. Within a short distance up the arroyo, the beds flatten substantially to a gentle dip of about 10 degrees. Soon after there is another major bend in the arroyo; along the northeast side is a rock pavement composed of a bedding surface of gray limestone with numerous intersecting fractures; on the opposite side are cliffs of red, tan, and gray rocks.

Figure 34. Lucero Arroyo.
A. View to the east-southeast from the parking spot with the Organ Mountains in the background, Doña Ana Mountains in the middle ground, and dark cliffs in Lucero Arroyo in the foreground; walk across the desert toward the dark cliffs to begin field trip.
B. Cobble conglomerate of the Plio-Pleistocene Camp Rice Formation.
C, D. Angular unconformity between gently tilted beds of the Eocene Palm Park Formation and nearly horizontal beds of the Camp Rice Formation looking north (C) and south (D).

E. Lahar (volcanic mudflow) of the Eocene Palm Park Formation.
F. Broad tidal channel in siltstone of the Permian Hueco Formation.
G. Two coiled gastropods (snails) and a long echinoid spine on bedding plane of a limestone of the Permian Hueco Formation.

These rocks belong to the Permian Hueco Formation and constitute the main theme of the field trip.

Exposed here are five rock types within the Hueco Formation: gray fossiliferous limestone, dark gray or green shale, tan dolomite, gray shale with light-colored nodules of calcite, and red or brown siltstone. The limestone and dark gray or green shale were deposited in a shallow sea, the dolomite and siltstone were deposited on a tide-dominated shoreline, and the shale with calcite nodules represents sedimentation on the land area above high tide. The nodules in the shale are of soil origin (caliche) and developed under relatively arid conditions. The siltstones have numerous sedimentary structures attesting to their mode of deposition, including broad tidal channels, crossbeds and ripples, and desiccation cracks (Figure 34f). The limestones contain marine invertebrate fossils, which are best seen at the end of the outcrop where the arroyo bends and limestone outcrops across the entire arroyo. The fossils, which are exposed on the gently dipping bedding surface, include large snails, long black shells of a clam similar to a modern razor clam, and tapering spines several inches long that were attached to echinoids, an animal similar to a sand dollar (Figure 34g).

The Hueco Formation was deposited along the western edge of the Orogrande basin, one of the major depositional basins produced during the Ancestral Rocky Mountains deformational event (Figure 17). The silt and clay were derived from mountains in northern New Mexico and deposited on a broad, gentle tidal flat. South of this site limestone and shale were deposited in a warm, shallow sea. Rise and fall of sea level on the time scale of a few hundred thousand years resulted in the shoreline moving back and forth over a distance of about ten miles. As a consequence, the Hueco Formation near Las Cruces consists of alternating beds of marine limestone and shale and tidal flat siltstone, dolomite, and shale, as can be seen here. Fluctuations in sea level probably were the result of the waxing and waning of continental glaciers that were situated at the south pole in Permian time.

Field Trip 9:
Robledo Mountains, near Las Cruces, NM

The primary purpose of this trip is to collect invertebrate marine fossils, although several other geologic features are also present. The area is reached by taking exit 9—Doña Ana from I-25 north of Las Cruces. If northbound, turn left (west) and go under overpass; if southbound, turn right. Proceed west for two miles on NM Highway 320, past stop sign and across railroad tracks. At junction with NM Highway 185, turn right (north) and then left, 0.4 miles later, onto Shalem Colony Trail. Two-tenths of a mile after crossing the bridge over the Rio Grande, turn right onto Rocky Acres Trail (CR DO-13). Just after the pavement ends, veer left at a Y-intersection; go left at another Y-intersection a short distance farther. Park in the gravel pit where the road splits into three roads that climb onto the mesa. The gravel is late Pleistocene in age and was deposited by streams draining the adjacent Robledo Mountains.

Walk up the gravel road on the far right (south) side. At the top of the hill, follow road to the right. After one hundred yards the road rises and forks; take the right fork that leads down into the arroyo. Where the road enters the arroyo bottom, turn right and walk down the arroyo about thirty yards, where there is an outcrop of beds of red siltstone and gray limestone of the Permian Hueco Formation tilted about 40 degrees to the east (Figure 35a, b). The Hueco is overlain by purple andesites of the Eocene Palm Park Formation. This outcrop marks the location of the East Robledo normal fault, which separates the Robledo Mountains from the Mesilla basin. The fault is just down-arroyo from the Palm Park Formation, and the tilt of the Hueco and Palm Park Formations is the result of movement on the fault. The fault became active in the latest Miocene or Pliocene Epochs, during the most recent phase of development of the Rio Grande rift.

Main Theme: Permo-Pennsylvanian Ancestral Rocky Mountains

Secondary Theme: Late Tertiary Rio Grande Rift

Access/Difficulty: Natural outcrops accessible by car and easy half-mile roundtrip hike on gravel road; longer, more difficult hikes optional

Figure 35. Robledo Mountains.

A., B. Beds of the Permian Hueco Formation tilted against the East Robledo fault.

C. Fossiliferous shale and limestone of the Hueco Formation.

Retrace the route to the gravel road and proceed west into the mountains. For the next two hundred yards there is very good fossil collecting on either side of the road. The most fossiliferous sites exist where erosion has exposed thin beds of brownish limestone and gray shale (Figure 35c). Look for individual fossils lying loose among the debris on the slopes. The fossils are in the Permian Hueco Formation, which was deposited in a shallow, tropical sea in the Orogrande basin (Figure 17). The fossils include several types of brachiopods, crinoid plates, snails, long black clam shells similar to modern razor clams, and echinoid plates and spines (cf. Figure 15). In order to preserve this site for future visitors, please collect only one of each type of fossil.

The fossil sites mark the end of the field trip. However, excellent hiking is available by following the road into the mountains to the place where it splits into three roads, with the middle one rising steeply. The road to the left is a particularly nice route into the mountains. The rocks encountered on this hike are limestone and dolomite of the Permian Hueco Formation and whitish rhyolite of the Bell Top Formation.

Field Trip 10:
Guadalupe Mountains, TX and NM

Main Theme: Permo-Pennsylvanian Ancestral Rocky Mountains

Access/Difficulty: Natural outcrops and roadcuts accessible by car; easy walks along road; more rigorous hikes available

Permian sedimentary rocks of the Guadalupe Mountains of west Texas and southern New Mexico are famous among geologists because of excellent exposures of reef limestone and dolomite that developed on the edge of the Delaware basin, as well as exposures of sandstone and shale deposited in deep water below the reef (Figure 17). Roadcuts of deep-water sandstone and shale exist along Highway 62-180 near Guadalupe Mountains National Park, whereas outcrops of the reef limestone and dolomite are seen west of White's City, New Mexico, on the road leading to Carlsbad Caverns. There is also an optional trip to McKittrick Canyon to see reef rocks. The stops described here are organized for travel eastward from El Paso to Carlsbad.

The parking spot for the first outcrop is a picnic area on Highway 62-180 on the steep rise below Guadalupe Pass. It is 4.7 miles east of the junction with Highway 54. From this spot there is an excellent view of limestone cliffs that make up El Capitan (Figure 36a). These cliffs represent gently dipping fore reef deposits composed of large blocks of rock broken from the reef and deposited as an apron on the seaward side of the reef. It is difficult to see from this angle that the fore reef deposits are tilted about 20 degrees toward the southeast, but the dip is evident from other vantage points farther to the east. The elevation difference between the limestone cliffs and the rest area is comparable to the difference in water depth between the reef and the deeper part of the Delaware basin—in late Permian time you would have been standing almost a thousand feet below sea level and looking up through the clear tropical sea at the reef.

From the rest area, walk downhill on the road about 0.5 miles toward the first outcrop, which is in the Permian Delaware Mountains Group (Figure 18). The outcrop can be separated into two parts: the upper part consists of medium to thick-bedded, light-colored sandstone and the lower part is thin-bedded gray siltstone and sandstone and dark gray shale. The light-colored sandstone was deposited in a broad submarine channel, as can be seen at the uphill end of the outcrop, where the sandstone-filled channel cuts down into the underlying sediment (Figure 36b). Deposition of individual beds of sandstone was probably by turbidity currents, which are chaotic mixtures of sediment and water that are pulled downslope by gravity into deep water. Turbidity currents are generated by slumps or earthquakes. The underlying sediment probably represents the finer grained fraction of turbidity currents that spilled out of the channel into the low areas between channels. Close examination of the gray siltstones and sandstones reveals ripples, indicating relatively low-velocity transportation of the sediment, consistent with an interchannel environment of deposition. Note also that the finer grained rocks have folds several feet in amplitude (Figure 36c). Because these folds are restricted to discrete layers and do not affect the rocks above and below, they formed shortly after the sediment was deposited and probably represent slumping toward the deeper part of the basin.

Leaving the rest stop and driving east toward the pass, there are large roadcuts of turbidite sandstone filling broad channels. The next official stop is 10.7 miles from the rest area, and 1.2 miles from Nickel Creek Station. It is a long roadcut at the top of a hill. Composed mostly of tan, fine-grained turbidite sandstone, there are several thin beds of gray conglomerate composed of pebble-sized clasts and fossils derived from the reef and probably also deposited by turbidity currents. The most impressive bed in the outcrop, however, consists of pebbles, cobbles, and boulders, measuring up to several feet across, of gray limestone "floating" in a matrix of tan sandstone (Figure 36d, e). The limestone clasts were derived from the reef, and the entire mass of sediment was probably transported and deposited by an underwater bouldery mudflow.

One mile northeast of stop 2 is the turnoff to McKittrick Canyon, which extends into fore reef and reef rocks of the Guadalupe Mountains. The Texas Bureau of Economic Geology has published a guide-

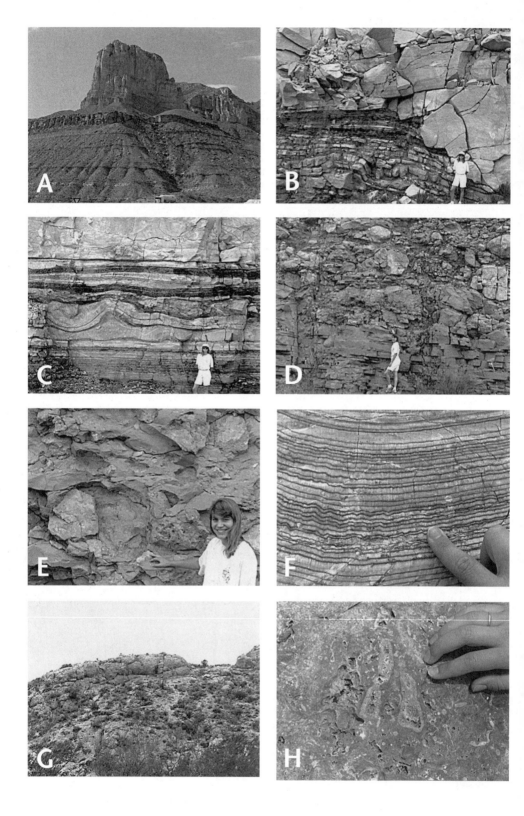

Figure 36. Guadalupe Mountains.
A. View from stop 1 of El Capitan, Guadalupe Mountains, Texas, representing Permian fore reef deposits. Thin ledges and weathered slopes below the peak are sandstone and shale deposited in deep water.
B. Submarine channel of the Permian Delaware Mountains Group at stop 1. Note how light-colored sandstone cuts into thin-bedded sandstone and shale.
C. Syndepositional folds in interchannel deposits at stop 1, probably formed by slumping of sediment toward deeper part of basin shortly after deposition.
D. Boulder conglomerate containing clasts of reef limestone up to a meter or more in diameter overlying thin-bedded sandstone at stop 2.
E. Close up of conglomerate shown in previous photo.
F. Thinly laminated gypsum and limestone of the Castile Formation at stop 3.
G. View from stop 4 at Permian reef core; note lack of bedding. To find fossils shown in the following photos, go across and halfway down the ridge in the lower left side of photo.
H, I. Fossil sponges making up the reef core at stop 4.
J. Well-bedded back reef sandstone and dolomite exposed on road to Carlsbad Caverns.
K. Teepee structure developed in back reef dolomite exposed in parking lot of Carlsbad Caverns.

book, available at the visitor's center, describing the rocks exposed in the canyon. It has outstanding pictures and diagrams of the reef complex at this location. Much of the information in the book is keyed to several self-guided trails.

Proceed on Highway 62-180 toward White's City. Approximately 15.2 miles from stop 2 is stop 3, a large roadcut of the Castile Formation, which is stratigraphically above the Delaware Mountains Group. The Castile Formation is composed of couplets a quarter of an inch thick of white to gray gypsum and dark brown limestone (Figure 36f). The couplets have been interpreted to represent annual deposits precipitated from the hypersaline Delaware basin. The water depth at the time of deposition of the Castile Formation was probably about one thousand feet. Also present in the layered gypsum-limestone beds are small folds due to slumping of the layers shortly after deposition and larger folds and beds of broken fragments that formed by a dissolving of the gypsum and the subsequent collapse of overlying beds.

Continue on Highway 62-180, turning left onto Highway 7, 13.6 miles northeast of the previous stop, into White's City. At the western end of White's City, park in front of the large sign marking the entrance to Carlsbad Caverns. Directly across the road and on the opposite side of a small arroyo is stop 4, exposures of reef limestone of the Permian Artesia Group (Figure 36g). The outcrop lacks obvious bedding and appears massive, which is due to the fact that it represents the core of the reef and is composed of fossils in their original growth positions. Unlike modern coral reefs, the Permian reef on the margin of the Delaware basin was composed of calcareous sponges and algae. The fossils are somewhat difficult to see in this outcrop because of recrystallization, but several large examples are visible, particularly on the back side of the outcrop (Figure 36h, i).

The remainder of the trip is on the road to Carlsbad Caverns National Park. There are several turnouts where rocks of the back reef can be seen. These rocks include well-bedded tan dolomite and yellowish sandstone deposited in shallow lagoons or on periodically emergent tidal flats (Figure 36j). Some of the dolomites contain quarter-sized spherical structures called pisolites, produced by soil processes. Also present in the dolomites are teepee structures, which are anticlines up to six feet in amplitude formed shortly after deposition of the sediment. A good outcrop of both pisolites and teepee structures is present on the edge of the first parking lot encountered when entering the visitor's center complex (Figure 36k). Remember, once you cross into the National Park you cannot collect samples or use a hammer on the rocks.

The caverns themselves are developed in rocks of the Permian reef complex, and the morphology of the caverns is related to the position within the reef. The first part of the cavern has rectangular rooms, because it is developed in the well-bedded back reef rocks. Eventually the cavern trail crosses into the reef core and the rooms become more irregular in their shape in response to the poor bedding of the reef core. Along the trail through the Big Room there is even a place where tilted beds of the fore reef can be seen.

Field Trip 11:
Cerro de Cristo Rey, El Paso, TX

Main Theme: Early Cretaceous Chihuahua Trough

Secondary Theme: Early Tertiary Igneous Intrusion

Access/Difficulty: Natural outcrops and roadcuts accessible by car; easy walks along road and on rocks

Cerro de Cristo Rey (also called Mount Cristo Rey), which straddles the border between Texas, New Mexico, and Mexico, has the most easily accessible outcrops of Lower Cretaceous sedimentary rocks in the region (Figure 37a). Consisting of a mixture of sandstone, siltstone, shale, and limestone, the Lower Cretaceous sediment was deposited in the Chihuahua Trough, which was a northwestward arm of the Gulf of Mexico (Figure 19). Mount Cristo Rey is cored by an early Tertiary intrusion of igneous rock (Figure 37b), and the Cretaceous sedimentary rocks have been uplifted and tilted steeply away from the center of the mountain. On this trip four of the Lower Cretaceous formations are encountered, several of which are fossiliferous. The igneous rock that cores the mountain is also available in outcrop.

It is prudent to note that the area around Cerro de Cristo Rey has a reputation for incidents involving muggings, especially on the steep trail leading to the statue of Jesus. I advise avoiding the trail to the statue altogether and recommend making the field trip with a group of at least four people.

From central El Paso, take Paisano Drive north and exit onto Highway 273 to Sunland Park; 0.3 miles after Highway 273 crosses the Rio Grande, turn left on the Mount Cristo Rey access road. If approaching the field trip site from the north, take I-10 East, exit at Sunland Park (exit 13), and turn right after exiting. After 0.4 miles turn left on Doniphan Road, go 0.5 miles, and turn right on Highway 498, following the signs for the racetrack. After 1.0 mile on Highway 498, turn left at the stop light and enter Highway 273. Go 0.6 miles on Highway 273 and turn right onto the Mount Cristo Rey access road. Once on the access road, refer to the map in Figure 37a. The outcrops to be examined occupy a rectangular area between the access road on the east and an arroyo on the west, and between the railroad tracks (Figure 37a). The exposures include natural outcrops and outcrops in inactive quarries. You may wish to park on the road leading to one of the inactive quarries. Do not go into the active quarry on the east side of the access road. As always, when examining rocks near abandoned quarries, be aware of the potential for falling rock from steep quarry walls.

The Cretaceous rocks in the area of the field trip dip from 50 to 80 degrees to the north. Consequently, the oldest rocks are exposed in the southern part of the area and they become progressively younger northward. The oldest formation is the Mesilla Valley Formation (Kmv; Figure 37a, c), which is exposed in the low slope just north of the southern railroad tracks. Composed of black shale and thin, tan siltstone and fine sandstone, the Mesilla Valley Formation was deposited in an offshore marine environment, in water whose depth ranged from ten feet to hundreds of feet. Burrows are common on the flat bedding planes of sandstone and siltstone, and it is also possible to find shell fragments of clams and oysters. One bed a foot thick near the road is composed almost exclusively of small oyster shells.

Holding up the small ridge north of the southern railroad tracks is medium bedded, brown sandstone of the Anapra Formation (Kan; Figure 37a, d). Crossbeds and thin, horizontal laminae are common in the Anapra Formation, but fossils are sparse, restricted to a few shell fragments. Near the top of the formation, black shale is interbedded with the sandstone. The Anapra Formation was probably deposited on the shoreline, either representing beach or delta environments, and when compared to the underlying Mesilla Valley Formation, indicates a relative lowering of sea level in the region. The origin of the sea level fall cannot be determined from one outcrop, but possibilities include a worldwide drop in sea level, a slowing down of local subsidence of the Chihuahua Trough, or an increase in sediment supply to the shoreline.

The inactive quarries on the west side of the access road are primarily positioned in the upper part of the Anapra Formation and the lower part of the Del Rio Formation (Figure 37e). The latter formation, which is composed mostly of black shale, is not well exposed in the area of the field trip. Because of its

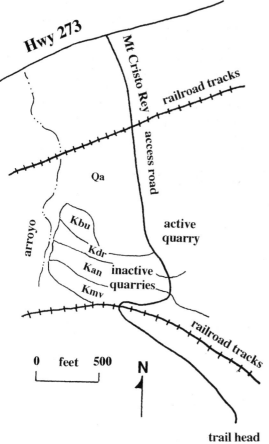

Figure 37. Cerro de Cristo Rey.
A. Map of the field-trip area on the north side of Mount Cristo Rey. Kmv=Cretaceous Mesilla Valley Formation; Kan=Cretaceous Anapra Formation; Kdr=Cretaceous Del Rio Formation; Kbu=Cretaceous Buda Formation; Qa=Quaternary alluvium.

B. View from the trail head of the statue of Jesus atop Mount Cristo Rey. Rocks adjacent to trail are the Muleros diorite.
C. Exposure of black shale and brown sandstone of the Mesilla Valley Formation.
D. View northward of ridge adjacent to southern railroad tracks. Lower slope is underlain by the Mesilla Valley Formation, whereas ridge crest is the base of the Anapra Formation.
E. Nearly vertical beds of the Anapra Formation exposed in an inactive quarry.
F. View to the northwest of light-colored outcrop of the Buda Formation.
G. Fossil gastropods (snails) in the Buda Formation.

fine grain size and the presence of marine fossils, mostly oysters and clams, the Del Rio Formation probably was deposited in an offshore marine environment, similar to the Mesilla Valley Formation. This interpretation indicates that after deposition of the Anapra Formation, sea level rose and the region was again inundated by a shallow sea.

The youngest formation exposed at the field-trip site is the Buda Formation, which is composed of very light gray limestone (Figure 37a, f). The Buda Formation is the most fossiliferous formation at this site, containing gastropods (snails), oysters, clams, corals, and echinoids. The fossils are easily seen in the

limestone itself (Figure 37g), and can also be found as individual specimens by combing through the pebble-sized weathered rubble that surrounds the limestone outcrops. The Buda Formation also was deposited in an offshore marine environment.

You may also wish to see, while on this field trip, the igneous rock that makes up the core of Mount Cristo Rey. Although officially designated the Muleros Andesite, I am more inclined to call it a diorite, because it is composed almost exclusively of minerals easily visible to the naked eye (Figure 2). These minerals include white feldspar, black, elongate hornblende, and black, platy biotite, some of which weathers to a shiny yellow or bronze color (but it is not gold). The Muleros Andesite/Diorite is one of a series of shallow intrusive igneous bodies about forty-seven million years old, making them roughly the same age as the Rubio Peak, Orejon, and Palm Park Formations (Figure 23). Others in the series include the Campus Andesite, near the Sun Bowl and the UTEP campus; and the Vado Andesite, which is visible from I-10 near the Vado exit. The Muleros Andesite/Diorite is well exposed on the trail to the statue of Christ (Figure 37b), as well as forming a small outcrop cutting across the Buda Limestone at the field-trip site proper.

Field Trip 12:
Mescal Canyon, Truth or Consequences, NM

Late Cretaceous sedimentary rocks exposed in Mescal Canyon near Truth or Consequences were deposited within and near the shoreline of the Western Interior Seaway, which stretched from the Arctic to the Gulf of Mexico (Figure 21). These outcrops are the best, most complete examples of Upper Cretaceous rocks in southern New Mexico. The rocks have been tilted about 40 degrees to the east during late Tertiary uplift of the Caballo Mountains. Here, it is possible to see evidence, through the interlayering of nonmarine and marine rocks, of two major changes in sea level. There is also an excellent fossil-collecting site near the end of the route.

Mescal Canyon can be reached from I-25 by taking either exit 75 or 79 to Truth or Consequences and proceeding into town. At the junction with Highway 51, turn east toward Elephant Butte Dam and Engle (at the writing of this book, this turn is at the only stop light in town). Approximately 3.1 miles after the turn, the road crosses a bridge over the Rio Grande; turn right onto a gravel road immediately after crossing the bridge. Follow the gravel road for 0.6 miles, making several bends, until it enters the large arroyo of Mescal Canyon. Turn left into the arroyo. At this point it is advisable for those in cars to park and walk the remainder of the route. Remember to leave room for vehicles to pass, because offroad vehicles can negotiate travel in the arroyo. From this parking spot, it is about a three-mile roundtrip hike along the arroyo (Figure 38a). About a mile up the arroyo, turn left (west) into a major side arroyo near a prominent outcrop of black shale and tan sandstone (Figure 38b).

During this trip five Upper Cretaceous formations will be encountered: Dakota, Mancos, Tres Hermanos, D-Cross, Gallup, and Crevasse Canyon (Figure 20). The main arroyo of Mescal Canyon roughly follows the Mancos Formation, and outcrops of thinly bedded black shale and tan siltstone can be seen on both

Main Theme: Late Cretaceous Western Interior Seaway

Access/Difficulty: Natural outcrops reached by car and a moderately difficult hike of three miles roundtrip in arroyo

sides of the arroyo. As you walk up the arroyo, the brown cliffs to the right (west) of the arroyo are sandstones of the Dakota Formation. Because the rocks are tilted to the east, the Dakota underlies the Mancos and is thus older. The cliffs to the left (east) side of the arroyo belong to the Tres Hermanos Formation, which is younger than the Mancos.

A good place to examine the Dakota Formation is an outcrop near the junction of the side arroyo with the main arroyo (labeled number 1 on the map in Figure 38a). The outcrop is a bedding surface of sandstone tilted toward the arroyo. At the south end of the outcrop is the contact between the Dakota sandstone and dark gray shale and siltstone of the Mancos Formation. The Dakota unconformably overlies Permian rocks and represents the oldest sediment deposited in the Late Cretaceous basin. The base of the formation, which is exposed halfway up the side of the ridge, was deposited by rivers, whereas the top, which is exposed in the arroyo, represents shoreline deposits. Thus, the Dakota shows evidence of the arrival of the Western Interior Seaway into southern New Mexico.

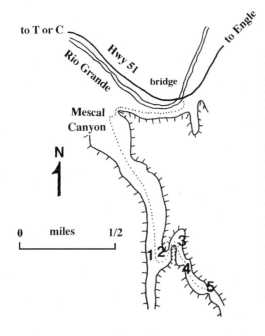

Figure 38. Mescal Canyon.
A. Route of field trip (dotted line).
B. Prominent outcrop on east side of Mescal Canyon that marks place where the field-trip route enters side arroyo. Lower beds are black shales and siltstones of the Mancos Formation; upper ledge is a sandstone at the base of the Tres Hermanos Formation.
C. Marine sandstone at the top of the Tres Hermanos Formation.
D. Thick, massive sandstones of the Gallup Formation at stop 5.
E. Burrows in the Gallup Formation made by a brine shrimp.
F. Fossil oysters in the middle of the Gallup Formation at stop 5.

Sea level continued to rise and deepen during deposition of the Mancos Formation, which is well exposed across the arroyo from the Dakota outcrop described in the previous paragraph (number 2, Figure 38a, b). At its deepest, the Mancos was deposited in hundreds of feet of water. The gray shale is sediment that slowly settled out of the seawater, and the thin beds of brown and tan siltstone and sandstone represent sediment carried seaward and deposited by storms. It is possible to find clam-shell fragments in the Mancos, as well as burrows along bedding planes.

During the latter part of deposition of the Mancos, the sea began to grow shallow and the shoreline built northeastward into the basin. Eventually the shoreline reached this area, depositing beach sandstones that overlie the Mancos (Figure 38b). These sandstones belong to the Tres Hermanos Formation. The shoreline continued to move seaward during deposition of the Tres Hermanos Formation and river sediment was deposited, which can be seen where the side arroyo makes a near right-angle bend (number 3, Figure 38a). The tan sandstones are crossbedded and were deposited in the river channels, whereas interbedded gray shales were deposited in adjacent floodplains. The final outcrop of the Tres Hermanos Formation is a light-colored sandstone that extends completely across the arroyo (number 4, Figure 38a, c). This sandstone has numerous burrows made by brine shrimp and was deposited near the shoreline as sea level once again began to rise and inundate this region. Sea level continued to rise, depositing dark gray, marine shales of the D-Cross Formation, which is best exposed on the ridge above the last Tres Hermanos sandstone. The D-Cross was deposited under conditions similar to the Mancos.

The last outcrop of the trip is a thick body of light-colored sandstone that extends across the arroyo (Figure 38d). This outcrop belongs to the Gallup Formation, which was deposited by a delta as sea level began to fall and the shoreline built out into the sea. The Gallup contains brine shrimp burrows, which are especially well exposed on the bedding surface just beyond the narrowest point of the arroyo (Figure 38e). The middle part of the Gallup, just beyond the narrows, was deposited in deltaic bays or lagoons and contains numerous, collectible oysters (Figure 38f). At the top of the Gallup is a dark gray bed of sandstone containing coalified roots; this is an ancient soil, or paleosol, and marks the end of the trip. All of the rocks beyond this point are river deposits of the Crevasse Canyon Formation. In northern New Mexico, sedimentary rocks equivalent to the Crevasse Canyon Formation have evidence of several more transgressions of the sea, but the shoreline did not return to our region after deposition of the Gallup Formation.

Cretaceous rocks like those in Mescal Canyon are economically important in the western United States and Canada, because they contain hydrocarbons (oil and natural gas). The hydrocarbons are generated in dark gray marine shales similar to the Mancos and D-Cross by the decomposition of tiny marine organisms. The hydrocarbons are expelled from the shales during burial and compaction and migrate into more permeable sandstones, like the Gallup, where they are trapped. In order to improve exploration strategies, oil-company geologists study outcrops like those in Mescal Canyon.

Field Trip 13:
San Diego Mountain, NM

Main Theme: Early Tertiary Laramide Deformation

Secondary Theme: Late Tertiary Rio Grande Rift

Access/Difficulty: Natural outcrops accessible by car; difficult four-mile roundtrip hike in arroyo and on outcrops

San Diego Mountain has one of the few easily accessible outcrops exhibiting early Tertiary Laramide deformation. It is one of the few places in the region where a Laramide fault is so well exposed that you can literally put your finger on it. Also present on this trip are river deposits of the ancestral Rio Grande and a well-exposed normal fault, both associated with the late Tertiary Rio Grande rift. Unless an offroad vehicle is utilized, the trip requires a rigorous four-mile roundtrip hike on a sandy road, in an arroyo, and on rocks.

The outcrops can be reached by taking I-25 to exit 32—Upham. If northbound, turn left and go under the overpass; if southbound, turn right. After crossing a cattle guard, the road becomes dirt and may have muddy patches following a rain; cars may not be able to cross large patches of standing water. The road is situated on the constructional top of sediment deposited by the ancestral Rio Grande. Beginning

Figure 39. San Diego Mountain.
A. View to the southeast of brown cliffs of sandstone of the Pliocene-early Pleistocene Camp Rice Formation. Cliffs end here, marking the point where unpaved road leads from main arroyo to narrow canyon.
B. View to south from main arroyo toward narrow canyon to be traversed on field trip.
C. View eastward of normal fault that uplifts San Diego Mountain.
D. Cross section of the view shown in photo "C"; TQc=Tertiary-Quaternary Camp Rice Formation; PCg=Precambrian granite and gneiss.
E. Laramide thrust fault.
F. Cross section at same scale as photo "E" of Laramide thrust fault; PCg=Precambrian granite and gneiss; PZsl=silicified Paleozoic limestone.

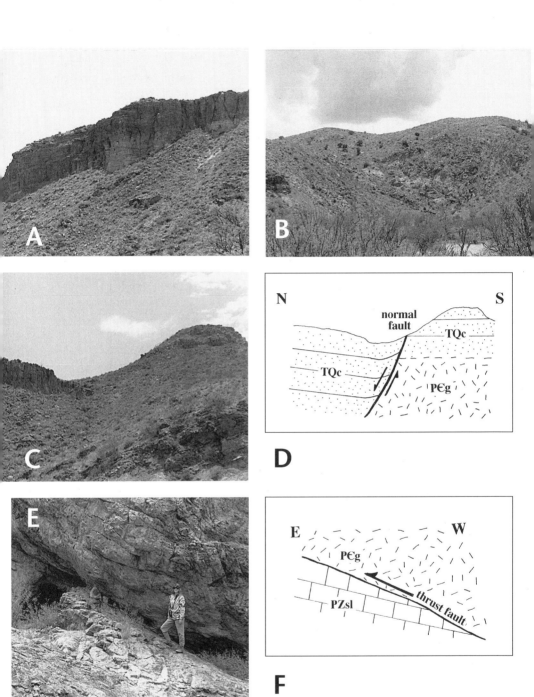

780,000 years ago, the Rio Grande began to erode, abandoning this surface and allowing a thick, white soil of calcite (caliche) to form, which can be seen in several ravines along the side of the road. About 4.6 miles from the interstate interchange, a barb-wire gate crosses the road; turn right before crossing the gate. Those in cars should park off to the side of the road before it begins to descend and walk the remainder of the route; offroad vehicles may proceed. Follow the main, sandy road, taking the left fork at the power line, where the road descends into the main arroyo. Turn right in the main arroyo. The brown cliffs on the left (south) side of the arroyo are river deposits of the Camp Rice Formation.

One mile after turning into the main arroyo the brown cliffs end (Figure 39a). Turn left onto a small unpaved road that heads into the hills (Figure 39b). Just before the road enters a narrow canyon is a good view of a normal fault of late Tertiary age. Precambrian rocks are on the footwall (upthrown side) and late Tertiary rocks of the Camp Rice Formation are on the hanging wall. Looking east, the fault can be seen to offset cliffs of the Camp Rice Formation, indicating that its most recent movement was after 780,000 years ago, which is quite recent geologically (Figure 39c, d).

Enter the narrow canyon. Exposed on the right (west side) are reddish, highly fractured Precambrian granites. About fifty yards beyond this point, the canyon is floored by rock: Precambrian granite on the right side and silicified Paleozoic limestone on the left side. Where the canyon makes a cliff with two small caves there is a thrust fault that dips about 30 degrees to the southwest and brings Precambrian granite up and over silicified Paleozoic limestone (Figure 39e, f). This fault was active during the Laramide deformational event in early Tertiary time and was one of the major faults of a large, northwest-trending mountain range that existed in our region at that time (Figure 22).

Exposed beyond this are outcrops of Precambrian granite, locally cross cut by veins of quartz and barite. Barite is an elongate, milky white mineral that is very heavy. Excellent samples of barite-bearing rock can be found as loose cobbles on the canyon floor.

Field Trip 14:
Faulkner Canyon, NM

Near the mouth of Faulkner Canyon, at the north end of the Robledo Mountains, are excellent exposures of early Tertiary rocks, along with dikes of younger volcanic rocks. The older rocks are green and purple andesites of the Palm Park Formation, which were associated with an Eocene volcanic arc that resulted from low-angle subduction of the Farallon plate beneath North America (Figures 22 and 23). Most of the rocks of the Palm Park Formation were deposited as volcanic mudflows, or lahars, although a few lava flows and ash flows are also present. The volcanoes responsible for Palm Park rocks were probably stratovolcanoes similar in appearance to Mount St. Helens. The dikes that crosscut the Palm Park Formation are also composed of andesite, but are a few million years younger. A plug of basalt is also present at the last stop. Field Trip 15 is very close to this trip and you may wish to do both trips in one day.

To reach Faulkner Canyon, take I-25 north of Las Cruces to exit 19—Radium Springs. If northbound, turn left and go over the overpass; if southbound, turn right. Follow Highway 157 west for 1.7 miles to the junction with NM Highway 185. Turn right on Highway 185. Approximately 1.7 miles from the junction turn left onto Faulkner Canyon Road (look for a green street sign). After a distance of 0.7 miles on Faulkner Canyon Road there is a gate across the road. Cars are advised not to go beyond this point, because the road is very sandy and cars may get stuck. In order to park safely off the road, it may be necessary to back up a short distance to find a pull out. From here, the trip requires a walk, entirely within the arroyo, of about five miles roundtrip. Those in offroad vehicles may drive the entire route, but remember to close the gate behind you.

The first outcrop encountered, 0.7 miles beyond the gate, is a steeply dipping dike on the right side of the arroyo, followed 0.1 miles farther on by an

Main Theme: Early Tertiary Laramide Volcanism

Access/Difficulty: Natural outcrops exposed along arroyo; without offroad vehicle, requires moderately difficult hike in arroyo of five miles roundtrip

Figure 40. Faulkner Canyon.
A. Unconformity between underlying green volcanic rocks of the Eocene Palm Park Formation and late Pleistocene gravel.
B. Cobble and boulders of andesite in a lahar of the Palm Park Formation.
C. Gently dipping beds of the Palm Park Formation; basal bed (person is standing on it) may be a lava flow, whereas pebbly beds above are lahars.
D. Spheroidal weathering of the basal bed pictured in photo "C."
E. Tilted beds of the Palm Park Formation; upper, thick beds are lahars; lower, thinner beds may be ash flows, lava flows, or stream-reworked volcanic sandstones.
F. Gently tilted beds of the Palm Park Formation cross cut by a more steeply dipping dike.
G. Plug of basalt.
H. Close-up of basalt plug, showing cooling joints.
I. Wall in arroyo made by steeply dipping dike.

outcrop of greenish lahars of the Palm Park Formation overlain by late Pleistocene gravels (Figure 40a). The contact between the Palm Park Formation and gravels represents an unconformity with over forty million years of erosion. About 0.2 miles beyond that are purple and green lahars of the Palm Park cut by a dike at the southern end of the outcrop (Figure 40b). About a mile farther up the arroyo, on the left side, is a large outcrop of gently dipping, dark purple Palm Park Formation (Figure 40c). The basal bed, which may be a lava flow, forms a ledge about six feet thick and consists of crystals of light green and white feldspar and black hornblende in a fine purple groundmass. Locally, this bed displays spheroidal weathering characteristic of many igneous rocks (Figure 40d). Above the basal bed are pebble and cobble lahars and several more ledges that may be lava flows or volcanic sandstones.

Two-tenths of a mile beyond the last outcrop is a large outcrop on the right side of the arroyo (Figure 40e). The uppermost bed is a thick cobble lahar, whereas the thinner beds beneath the lahar consist of lava flows, volcanic sandstones, and perhaps ash-flow tuffs. The exact origin of the lower beds is equivocal without examination under the microscope. A few hundred yards farther, on the left side of the arroyo, is an excellent example of a dike cutting across gently dipping beds of the Palm Park (Figure 40f). The dike is a hard, greenish gray andesite with no visible crystals. Keep in mind the fundamental principle of igneous rocks: the cross cutting rock (the dike in this case) is younger than the rocks it cross cuts (the Palm Park Formation here).

A few hundred yards beyond the last outcrop is an abandoned ranch; follow the arroyo to the right of the ranch. Three-tenths of a mile beyond the ranch is a large body of dense, black to dark gray basalt, displaying complex cooling joints (Figures 40g, h). The age of the basalt has not been determined and it may be coeval to the dikes that cross cut the Palm Park. A tenth of a mile beyond the basalt, on the right side of the arroyo, is an excellent outcrop of a near vertical dike that resembles a wall (Figure 40i). The trip ends here. Although the road continues for many miles, the geology does not change very much.

Field Trip 15:
Cedar Hills, NM

This field trip follows the same route as the previous trip to Faulkner Canyon (Field Trip 14) and you may wish to do both trips in the same day. Turn left off NM Highway 185 onto a gravel road 0.8 miles north of the junction with Faulkner Canyon Road. The road is called Fred Huff Road, but it does not have a sign. There is a cattle guard immediately after the turn and then the road ascends to a mesa. After 2.1 miles the gravel road descends into a canyon and crosses a small arroyo. One mile after crossing the arroyo, turn right onto a small ranch road and, 0.1 mile farther, park in a turnout on the left side of the road, before it descends into the valley. From this vantage point it is possible to look northwestward into the canyon below and see a dark, rounded knob in the middle distance, which is one of the outcrops to be examined (Figure 41a).

Main Theme: Mid Tertiary Volcanism

Access/Difficulty: Natural outcrops reached by car; moderately difficult hike on gravel road and in arroyo; longer optional hike available

The volcanic rocks exposed in this trip are middle Tertiary (Oligocene Epoch) in age and formed at the same time as the rhyolite calderas of the Organ Mountains, Black Range, and elsewhere in southwestern New Mexico and west Texas. The volcanic rocks exposed here are related to a series of more than two dozen flow-banded rhyolite domes, called the Cedar Hills Vent Zone, which were probably injected along a fault (Figure 24). In this trip you will see several different types of volcanic rock associated with the vent zone.

Walk down the road toward the canyon floor. Where the road enters the arroyo, turn right into the arroyo. After about thirty yards, outcrops of dark volcanic rock are exposed along the side of the arroyo (Figure 41b). These rocks are composed of angular pieces of black, shiny obsidian containing scattered white crystals of quartz and feldspar. Obsidian is a volcanic glass, representing rhyolitic lava that solidified so quickly that only a few minerals were able to form. Because it is a glass and contains so few primary minerals, obsidian does not follow

Figure 41. Cedar Hills Vent Zone.
A. View northwestward from parking spot toward flow-banded rhyolite dome in the middle distance.
B. Volcanic lava flow made of fragments of black, glassy obsidian.
C. Flow banding in the rhyolite dome pictured in photo "A."
D. Rhyolitic volcanic sand and gravel redeposited by sedimentary processes, including shallow, high-energy stream flows and mudflows.
E. Thick deposit of ash-flow tuff, showing indistinct vertical cooling joints.

the volcanic-rock classification scheme that relates rock color to composition. The obsidian erupted as a viscous lava that was very resistant to flow. As it moved sluggishly along, parts of the flow solidified and then were broken up and reincorporated into the lava. This process accounts for the angular pieces that make up the obsidian outcrop.

Continue down the arroyo about twenty yards to outcrops at the base of the steep knob. These white, pink, and light gray rocks exhibiting distinct layering are flow-banded rhyolite (Figure 41c). On a fresh surface it is possible to see a few glassy crystals of quartz and clear, shiny crystals of feldspar set in a fine matrix. The layering (flow banding) is the result of sluggish flow of the rhyolite magma, which was injected upward into the surrounding rocks as steep cylindrical or funnel-shaped bodies and accounts for the near vertical flow banding. The flow-banded rhyolite domes are analogous to blisters that do not penetrate the outer layer of the skin, but cause a bulge in it. Occasionally, lava did erupt from the domes, forming ash falls, ash flows, and lava flows, including the obsidian flow observed at the previous outcrop. As you follow the arroyo around the bend, you are standing in the middle of one of the flow-banded rhyolite domes.

After examining the flow-banded rhyolite, retrace the route in the arroyo. Go about two hundred yards beyond where the road crosses the arroyo to an outcrop of whitish sandstone and conglomerate (Figure 41d). These rocks represent mostly ash-fall material that was reworked by mudflows and small, ephemeral streams. Although composed entirely of volcanic debris, these rocks were ultimately deposited by sedimentary processes. Owing to their mixed origin, rocks such as these are called volcaniclastic rocks.

Continue up the arroyo about 150 yards to outcrops of another flow-banded rhyolite dome. Flow banding is not as conspicuous in this outcrop, and the rhyolite is broken into angular blocks, probably by the same process described previously for the obsidian. This is the end of the trip; retrace the route to return to your vehicle.

An optional hike of several miles is available from this site. Follow the road past the arroyo, over the low ridge, and into the large arroyo of Foster Canyon. Upon entering the large arroyo, turn left. At a right-angle bend in the arroyo, reddish brown cliffs appear to the west-southwest. These cliffs are composed of ash-flow tuff, which probably was derived from the Cedar Hills Vent Zone. A few hundred yards later the road leaves the arroyo and moves up onto whitish beds, which represent ash-fall tuffs and ash-fall material reworked by sedimentary processes. There is an excellent view of the ash-flow tuff making up the surrounding cliffs (Figure 41e).

Field Trip 16:
Florida Mountains, near Deming, NM

The Florida and Little Florida Mountains, located about fifteen miles southeast of Deming, New Mexico, contain beautifully exposed early and middle Tertiary volcanic rocks, which are accessible at two state parks, Rock Hound and Spring Canyon, situated only a few miles apart. The route to the parks is well marked with signs that begin at exit 82A of I-10 in Deming. Both parks have picnic spots and require a day use fee, and Rock Hound State Park has overnight camping facilities.

The volcanic rocks exposed at Rock Hound State Park are Oligocene in age and erupted at roughly the same time as the Organ Mountains and Emory calderas (see Field Trips 17 and 18). The rocks were subsequently uplifted along late Tertiary normal faults associated with the Rio Grande rift. The large cliffs overlooking the state park are composed of reddish flow-banded rhyolite that formed as lava flows, shallow intrusive domes, and cross cutting dikes (Figure 42a). The rhyolites commonly exhibit a texture of angular, broken pieces produced as parts of the sticky magma crystallized and then were incorporated back into the magma. Less well exposed, but present as weathered debris in the arroyos, are orangish red ash-flow tuff and black and brown obsidian, a rhyolitic volcanic glass. Some of the brown glass, in particular, exhibits minute, curved, concentric cracks that resulted from expansion as the rock gained water; such rocks are called perlites. What makes Rock Hound State Park unique is that visitors are encouraged to collect up to a maximum of fifteen pounds of rocks and gems. Especially common is jasper, a red form of fine-grained quartz, and geodes, which represent cavities filled with crystals of quartz and perhaps other, unusual minerals.

Spring Canyon State Park is underlain entirely by the late Eocene Rubio Peak Formation (Figure 23).

Main Themes: Early and Mid Tertiary Volcanism

Access/Difficulty: Natural outcrops in state parks; accessible by car; easy hikes on trails

Figure 42. Florida Mountains.
A. View eastward of Little Florida Mountains at Rock Hound State Park. Rocky cliffs are flow-banded rhyolite.
B. Gently tilted volcanic rocks of the Eocene Rubio Peak Formation at Spring Canyon State Park.

The volcanic rocks are gently tilted to the northeast and weather into a variety of scenic landforms (Figure 42b). Composed of purple, gray, and green andesite, most of the Rubio Peak Formation consists of bouldery lahars, although a few lava flows and volcanic sandstones are present as well. Close examination of the andesite reveals white crystals of feldspar and elongate black hornblende surrounded by a fine matrix. Locally, the andesite is altered to a light green, fine-grained mineral called epidote. Surrounded by thousands of feet of andesite at Spring Canyon Park, it is possible to appreciate the great volume of volcanic rock erupted from the early Tertiary stratovolcanoes. At that time our region may have resembled the Cascade Range of northern California, Oregon, and Washington.

Field Trip 17:
Dripping Springs State Park and San Augustin Pass, near Las Cruces, NM

Main Theme: Mid Tertiary Calderas

Access/Difficulty: Natural outcrops accessible by car and moderately difficult hikes on trails

The Organ Mountains caldera is one of a series of about a dozen calderas active in southwestern New Mexico and west Texas in Oligocene time (Figure 24). Erupting between 36.2 and 35.4 million years ago, the Organ Mountains caldera is the oldest in New Mexico. Like so many of the other calderas, late Tertiary normal faulting has segmented portions of the Organ Mountains caldera, such that it no longer has its original form. However, up to ten thousand feet of ash-flow tuff that was trapped within the caldera walls are exposed in the Organ Mountains. The tuffs are separated into four mappable formations, each of which includes numerous individual ash-flow events (Figure 23). The character of the tuffs reflects processes acting in the magma chamber, particularly the sinking of early crystallized minerals to the bottom of the chamber. The oldest tuffs (Cueva and Cox Ranch) represent lava from the top of the magma chamber and thus have only a small number of early formed crystals. In contrast, the younger eruptions tapped the more crystal-rich lower parts of the magma chamber and the resulting tuffs have a higher percentage of visible minerals of quartz, feldspar, and mica. The Organ Mountains caldera is also unique because the plutonic rocks that crystallized in the magma chamber beneath the volcano are exposed. Both the extrusive rhyolite tuffs and the intrusive granite are encountered on this trip.

The trip begins with examination of the volcanic rocks at Dripping Springs State Park near Las Cruces, which can be reached by taking the University Avenue exit (Exit 1) off I-25. If northbound, turn right; if southbound, turn left and cross the overpass. Go east on University Avenue a total of ten miles, following the signs to Dripping Springs State Park. Once you enter the park, remember that you are not allowed to use your hammer, and sampling is prohibited. Although numerous trails exist within the park, two of the best for understanding the geology are the Cueva Rock trail and the Dripping Springs trail. Cueva Rock is made of the Cueva ash-flow tuff, which at this location contains abundant pieces of whitish pumice, as well as angular, dark fragments of rock scraped off the walls of the magma chamber or derived from the walls of the caldera and incorporated into the erupting ash flow. This type of volcanic rock is called a "lithic tuff." Although it is difficult to see the original volcanic layering, the beds of tuff are tilted about 50 degrees to the southwest (Figure 43a).

The first outcrops of volcanic rock encountered on the Dripping Springs trail are reddish brown, very dense crystal-poor ash-flow tuff with flattened whitish or light gray pieces of pumice belonging to the Cox Ranch Formation. The trail then crosses a large arroyo and heads up into the mountains. The arroyo roughly marks the position of the Modoc fault, an arcuate, down-to-the-east normal fault that is interpreted to have formed when a huge mass of volcanic material fell into the underlying magma chamber. Once past the Modoc fault, all the volcanic rocks surrounding the trail are part of the Squaw Mountain tuff and exist as beds dipping between 30 and 70 degrees to the west (Figure 43b). Conspicuously displayed in the tuffs are cooling fractures oriented at a high angle to the original layering (Figure 43c). Similar in appearance to the tuff of the Cox Ranch Formation, the Squaw Mountain tuff is reddish brown and dense, and has light-colored pieces of pumice that have been flattened to such a degree that the rock almost resembles a flow-banded rhyolite. Unlike the tuff of Cox Ranch, the Squaw Mountain tuff generally has visible crystals of glassy quartz, shiny white feldspars, and black biotite. By the time you've reached the end of the trail, you should begin to appreciate the great volume of ash-flow tuff that accumulated within the caldera.

Retrace the route from Dripping Springs State Park westward, turning right (north) on Baylor Canyon Road. After 6.8 miles on this gravel road, turn right onto U.S. Highway 70. Just past San Augustin Pass, 3.2 miles after turning onto Highway 70, turn right into a large, paved turnout. At the far western end of the turnout is a trail that leads to a covered picnic

site. Go beyond the picnic site, through the wire fence, to the far end of the trail, where there are several natural outcrops of rock. These outcrops are part of the Organ Mountains granite and represent the magma that slowly crystallized beneath the Organ Mountains caldera. Two phases of the granite are present. The coarse variety consists of large (quarter and dime-sized) minerals of white feldspar, gray, glassy quartz, black, equidimensional biotite, and black, elongate hornblende (Figure 43d). The finer variety has a sparkly appearance and is composed of gray quartz and white feldspar. Using the igneous rock classification of Figure 2, these rocks would be called granite, because they are light in color and consist entirely of minerals visible to the naked eye. Several larger-scale features of granite can be seen as well. Granite tends to weather into rounded boulders, as can be seen in the outcrops at the end of the trail, and as smooth sheets, as can be seen in the far distance to the southeast in a conical peak called Sugarloaf Peak (Figure 43e). The adjacent hill to the south-southwest also exhibits cooling joints (Figure 43f).

Although the trip ends here, an optional site where the Organ Mountains granite can be seen is Aguirre Springs State Park, which can be reached by traveling east on Highway 70 for one mile and turning right onto the park road. Like Dripping Springs Park, Aguirre Springs requires a day use fee.

Figure 43. Organ Mountains Caldera.
A. Cueva Rock, composed of the Oligocene Cueva ash-flow tuff.
B. Tilted beds of the Squaw Mountain ash-flow tuff as viewed from the Dripping Springs Trail.
C. Cooling joints in the Squaw Mountain ash-flow tuff.
D. Coarsely crystalline phase of the Organ Mountains granite.
E. Sugarloaf Peak, an example of weathering of granite into smooth, sheetlike surfaces.
F. Vertical cooling joints in the Organ Mountains granite.

Field Trip 18:
Black Range, NM

Main Theme: Mid Tertiary Calderas

Secondary Theme: Rio Grande Rift

Access/Difficulty: Roadcuts and natural outcrops accessible by car and easy walks along road

The Emory caldera was among about a dozen rhyolitic volcanic centers that erupted in the Oligocene Epoch in southwestern New Mexico and west Texas (Figure 24). The Emory caldera, like so many of the others, does not retain its original volcanic shape, because late Tertiary normal faults associated with the Rio Grande rift segmented the volcanic edifice into discrete fault blocks, uplifting some parts of the caldera and downdropping others. Despite this modification, the caldera rocks are well exposed in the Black Range and surrounding areas and the original nature of the caldera can be reconstructed.

Many of the volcanic rocks associated with the Emory caldera, along with several younger and older rocks exposed in late Tertiary fault blocks, are exposed on a transect of the Animas Mountains and Black Range on NM Highway 152 (formerly NM Highway 90). Highway 152 begins at exit 63 (Hillsboro) of I-25. If northbound, turn left and go under the overpass; if southbound, turn right. The road quickly rises to a mesa, past outcrops of Pliocene and early Pleistocene conglomerate and sandstone derived from the Black Range and deposited in the Palomas basin. The top of the mesa, where the road flattens, is the constructional top of the wedge of sediment deposited in the Palomas basin. This surface was abandoned about 780,000 years ago, when the Rio Grande and its tributaries began to cut down through its own pile of sediment.

About 12.7 miles west of the junction of NM 152, just beyond a dip in the road, is a roadcut of light tan volcanic rock that constitutes stop 1 of the trip (Figure 44a). This rhyolite ash-flow tuff belongs to the Kneeling Nun Formation (Figure 23) and erupted from the Emory caldera about thirty-five million years ago. The tuff is light tan in color and consists of numerous crystals of glassy quartz, shiny gray feldspar, and black biotite in a groundmass of fine ash. Also present are large, oblong pieces of whitish pumice (Figure 44b). This outcrop represents an ash flow that escaped the caldera and spread across the countryside, at this spot about ten miles from the inferred eastern edge of the caldera. You may wish to collect a small sample and compare it to other samples of the Kneeling Nun tuff encountered later on the trip.

From stop 1 to the town of Hillsboro the road crosses the Animas Mountains, a late Tertiary normal fault block, whose main border fault is on the west side. The Animas Mountains is one of a series of fault blocks that exist between stop 1 and Emory Pass in the Black Range in which Paleozoic sedimentary rocks and Cenozoic volcanic rocks are exposed. The pinkish rocks exposed in the low hills around Hillsboro are Miocene conglomerates that were deposited during the early stage of block faulting and were subsequently faulted themselves. The flat mesa north of Hillsboro is underlain by a basalt lava flow that erupted about four million years ago (Figure 44c).

Just after entering the outskirts of Hillsboro, the highway crosses a bridge over Percha Creek. The second stop of the trip is 2.5 miles from this bridge, at a large pull out on the opposite side of the road. Exposed here in a narrow canyon are basalt lava flows of the Uvas Formation, which erupted twenty-eight million years ago (Figure 23). The outcrop exhibits cooling joints (Figure 44d), and contains a few white to glassy feldspar crystals in a dark, fine groundmass.

For stop 3, pull off to the left side of the road, 1.1 miles from the last stop, at the east end of a long outcrop of gently east-tilted gray shales and brown sandstones unconformably overlain by Quaternary gravels (Figure 44e). The tilted shales and sandstones were deposited in an Oligocene lake that developed after activity in the Emory caldera, but before eruption of the Uvas basalt. If you dig into the shale exposed near the east end of the roadcut, you may find small plant fossils, mostly pine needles.

The trip returns to the main theme of the Emory caldera at stop 4, which is 1.5 miles from stop 3. Pull off to a gravel turnout on the right side of the road and examine the series of roadcuts that extend for about two hundred yards to the west. The east end of the outcrop, across from the parking spot, con-

Figure 44. Emory caldera, Black Range.
A. The Oligocene Kneeling Nun Tuff, at stop 1, representing a thin ash-flow deposit that escaped the walls of the Emory caldera.
B. Close-up of ash-flow tuff at stop 1, showing white pumice lumps.
C. Mesa above the town of Hillsboro is held up by Pliocene basalt lava flow.
D. Cooling joints (columnar joints) of the Oligocene Uvas basalt at stop 2.
E. Gently tilted Oligocene shale deposited in a lake and overlain by Pleistocene gravel. The contact is an unconformity.
F. White rhyolite of probable ash-fall origin at stop 4.

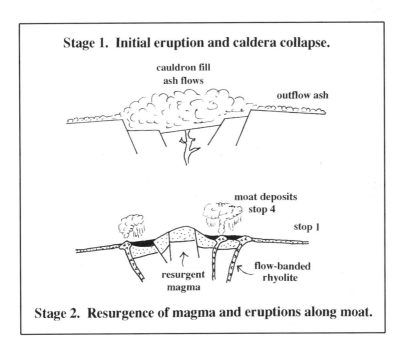

G. Flow-banded rhyolite at stop 4.
H. Two-stage model for development of the Emory caldera, adapted from Elston, W. E., Seager, W. R., and Clemons, R. E., 1975, New Mexico Geological Society, Guidebook 26, p. 283–292.
I. Kneeling Nun tuff that filled the Emory caldera.

sists of white, tan, and pink rhyolite that probably represents ash-fall deposition (Figure 44f). Exposed in the central part of the series of roadcuts is flow-banded rhyolite, one outcrop of which forms a small fold (Figure 44g). At the far western end, in an outcrop that is steep and potentially dangerous due to falling rock, is more flow-banded rhyolite that intrudes white tuff and dark volcanic glass. The volcanic rocks exposed at stop 4 illustrate the complex history of caldera eruptions. Many calderas, including the Emory caldera, develop in two major stages (Figure 44h). The first stage involves initial eruption and collapse of the volcano into a broad caldera. During the second stage, magma surges upward beneath the caldera, doming the previously erupted volcanic tuffs and emplacing flow-banded rhyolite domes and ash-fall deposits along the edge of the dome, an area referred to as a moat. The volcanic rocks exposed at stop 4 represent moat deposits formed during the second stage of the Emory caldera, and thus define the eastern edge of the caldera.

The next stop is at the Emory Pass Vista Point, which is reached by turning right from the main highway twelve miles from stop 4. Between stop 4 and the vista point, there are numerous roadcuts of gray limestone of Paleozoic age, some of which have been metamorphosed to marble by the heat and fluids that emanated from the Emory caldera, black shale of the Devonian Percha Formation, red sandstone and shale of the Permian Abo Formation, purple andesite lahars of the Eocene Rubio Peak Formation, and tan and brown-weathering rhyolites related to the Emory cauldron. These rocks are repeated numerous times as a result of late Tertiary faulting. See if you can identify these various formations by their color as you view them from the highway. From the Emory Pass Vista Point you can see Highway 152 and the towns of Hillsboro and Kingston, as well as several mountain ranges produced by late Tertiary block faulting.

After leaving the vista point, turn right (west) onto Highway 152. For the next four miles there are outcrops of the same rocks encountered during the drive up to Emory Pass. Beyond this distance, however, all of the roadcuts for 8.6 miles are Kneeling Nun tuff that was trapped inside the Emory cauldron. As you drive along and see outcrop after outcrop of the white, tan, and pink tuff, you should begin to appreciate the massive volume of volcanic debris erupted by the Emory cauldron. There are numerous pull outs along the road; stop at several of your choice and examine the rocks. The Kneeling Nun tuff along this transect is similar in appearance to stop 1, containing numerous crystals of quartz, feldspar, and biotite in an ash matrix, as well as white blebs and streaks up to fist size of pumice. Most of the outcrops display steeply dipping cooling fractures (Figure 44i), and a few outcrops show spheroidal weathering. Several outcrops of the tuff have boulder-sized chunks of limestone, red sandstone, or purple andesite that represent blocks which were brought up with the magma or fell into the caldera from its steep sides. Locally within the Black Range, some of these blocks are as large as houses.

About 12.6 miles from the Emory Pass Vista Point, the Kneeling Nun tuff is overlain by dark basalt, which in turn is overlain by the Gila Conglomerate. From this point it is only a view miles to the town of San Lorenzo.

Field Trip 19:
Apache Canyon, Caballo Mountains, NM

Main Theme: Late Tertiary Rio Grande Rift

Secondary Theme: Early Tertiary Laramide Deformation

Access/Difficulty: Natural outcrops reached by car and a moderately difficult 2.5 mile roundtrip hike in an arroyo

Late Tertiary crustal extension in the southern Rio Grande rift took place in at least two phases, the initial phase in latest Oligocene and Miocene Epochs and the later phase in latest Miocene, Pliocene, Pleistocene and Holocene Epochs. Apache Canyon in the Caballo Mountains is a good place to study both phases of the Rio Grande rift, because sedimentary rocks deposited during the initial phase have been uplifted and exposed by faults developed during the later phase. Moreover, incision of the arroyo in Apache Canyon has exposed the Plio-Pleistocene sediment deposited during the later phase of faulting and basin subsidence. In addition to rift-related rocks, Apache Canyon also has outcrops of an angular unconformity involving sedimentary rocks deposited during the early Tertiary Laramide mountain building event.

Apache Canyon is reached by taking the Caballo-Percha State Parks exit (exit 59) off of I-25. If northbound, turn right; if southbound, turn left and cross the overpass. A few hundred yards northeast of the overpass, turn right toward Caballo Dam. Continue straight and about 0.9 miles farther cross Caballo Dam on a one-lane bridge. After crossing the bridge, the road splits into three, the road on the right going down toward the river and the left road leading to a sign describing day-use fees; take the middle road that leads up a small rise and bends to the right. About 0.5 miles from the bend, the road descends into a valley and intersects the broad arroyo of Apache Canyon. If you are in a car, it is best to park here and walk eastward up the broad arroyo; offroad vehicles can drive on the road in the arroyo. From this parking spot it is about a 2.5-mile-roundtrip walk in the arroyo.

The horizontal sedimentary rocks exposed along the walls of Apache Canyon, both 0.5 miles west and

0.7 miles east of the parking spot, belong to the Pliocene-early Pleistocene Palomas Formation (Figure 45a) and represent the sedimentary record of the most recent phase of deformation in the Rio Grande rift (Figure 23). The crossbedded sand exposed along the gravel road where it descends into Apache Canyon was deposited by the ancestral Rio Grande (Figure 45b), whereas coarse conglomerates above the river sand and in outcrops upcanyon from the parking spot were deposited in shallow, ephemeral streams draining the Caballo Mountains. Some of these outcrops are at arroyo level and can be examined closely while walking up the broad arroyo of Apache Canyon.

Approximately 0.7 miles from the parking spot, the arroyo crosses the trace of a large normal fault, called the Red Hills fault, which uplifts the southern portion of the Caballo Mountains. To the right (south) are steeply dipping limestones of the Pennsylvanian Magdalena Group (Figure 45c), and on the left is the purple Eocene Palm Park Formation and the steeply dipping latest Oligocene-Miocene Hayner Ranch Formation, which will be examined later on the trip. The Red Hills fault is a late-rift structure, which developed within an early rift basin and uplifted some of the sediment deposited in the older basin.

On the left side of the arroyo, 0.2 miles upcanyon from the Red Hills fault, are red conglomerates and shales of the Eocene Love Ranch Formation. Continue walking up the arroyo another 0.3 miles to a place where the road leaves the arroyo and begins to climb a small hill (Figure 45d). Stay in the arroyo at this point and follow it a few hundred yards as it bends to the right. The red conglomerates and shales exposed here also belong to the Eocene Love Ranch Formation and were deposited in a basin situated northeast of a large, northwest-trending, basement cored, thrust-fault bounded uplift. The tremendous size of some of the boulders in the conglomerates attests to deposition very near the mountain front (Figure 45e). The cobbles and boulders making up the Love Ranch conglomerates include clasts of Paleozoic limestone, red Permian siltstone, Precambrian granite, and green Cretaceous andesite. Also exposed here is an angular unconformity between the Penn-

Figure 45. Apache Canyon, Caballo Mountains.
A. View eastward into Apache Canyon. Pliocene river sand deposited by the ancestral Rio Grande is exposed along road in foreground. Cliffs in left middle ground are Pliocene and early Pleistocene conglomerates.
B. Crossbedded sand of the Pliocene-early Pleistocene Palomas Formation deposited by the ancestral Rio Grande.
C. Steeply dipping beds of the Pennsylvanian Magdalena Group upturned against the Red Hills fault.
D. Outcrop of Eocene Love Ranch Formation at turnaround point of field-trip route.
E. Boulders of Paleozoic limestone and dolomite in the Eocene Love Ranch Formation, located fifty yards up the arroyo from previous photo.
F. Angular unconformity between limestones of the Pennslvanian Magdalena Group below and red shale and conglomerate of the Eocene Love Ranch Formation above, located a few yards up the arroyo from previous photo.
G. Outcrops of the Oligocene Thurman Formation as seen from Apache Canyon.
H. Thick beds of the Thurman Formation deposited as mudflows.
I. Thin beds of the Thurman Formation deposited by shallow streams.
J. Beds of the latest Oligocene-Miocene Hayner Ranch Formation.
K. Close up of cobble conglomerates of the Hayner Ranch Formation.

sylvanian Magdalena Group and the Eocene Love Ranch Formation (Figure 45f). Note that the Pennsylvanian rocks are dipping more steeply than the Eocene rocks, indicating that the Pennsylvanian rocks were tilted prior to deposition of the Eocene Love Ranch conglomerates, and then both formations were tilted to their present position during the later phase of the Rio Grande rift.

After examining the Love Ranch Formation, turn around and retrace the route down the arroyo. At the point where the red beds of the Love Ranch Formation were first encountered (0.3 miles from turnaround point), turn right (north) and walk up a small arroyo/road toward a prominent outcrop of white rocks (Figure 45g). These rocks, which are well exposed and accessible in the small arroyo, are part of the late Oligocene Thurman Formation (Figure 23). They are composed primarily of minerals, glass shards, and fine pumice that were exploded out of the Mount Withington caldera about 27.5 million years ago (Figure 24). The ash-fall material mantled the countryside and then was redeposited by ephemeral streams and mudflows. The stream deposits are fairly well sorted and thin bedded, whereas the mudflows are thick bedded and have a few scattered cobbles and numerous white pebbles of pumice surrounded by a sandy matrix (Figure 45h, i). Also present at this outcrop are two thin beds of white ash that were not reworked by sedimentary processes. Block faulting associated with the early phase of the Rio Grande rift had already begun in the region at the time of deposition of the Thurman Formation, but it is unlikely that the Caballo Mountains had begun to be uplifted at this time, because there is no evidence of coarse sediment derived from the Caballo Mountains in the Thurman Formation.

After examining the Thurman Formation, retrace the route back down the small arroyo. About twenty yards before reaching the barb-wire fence, turn right toward a badly eroded road that rises up the side of a hill. Just past the crest of the hill is a good view to the right (north) of tilted gray beds of the late Oligocene-early Miocene Hayner Ranch Formation, which conformably overlies the Thurman Formation (Figure 23). Even from this vantage point it is possible to see that the Hayner Ranch Formation is composed of conglomerates and interbedded sandstones (Figure 45j). These rocks can be examined more closely by continuing to follow the eroded road into the next arroyo (Figure 45k). The Hayner Ranch Formation was deposited in a basin directly adjacent to the rising Caballo Mountains during the early phase of the Rio Grande rift. When the Red Hills fault formed, the rocks of the Hayner Ranch Formation were uplifted and tilted to their present position. This is an excellent example of how rift sediment can be subsequently involved in faulting.

Walk down the arroyo to the south, toward the main arroyo of Apache Canyon. Turn right and proceed down the canyon to the parking spot. A short distance after reentering the main arroyo, you will cross the Red Hills fault and reenter the Pliocene and Pleistocene Palomas basin.

Field Trip 20:
Rincon, NM

Main Theme: Late Tertiary Rio Grande Rift

Access/Difficulty: Natural exposures accessible by car and easy hikes

The two-phase history of the late Tertiary Rio Grande rift is illustrated in outcrops near the village of Rincon, thirty-two miles north of Las Cruces. In latest Oligocene and Miocene time, this area was a depositional basin, in which several thousand feet of sedimentary rocks were deposited. These red rocks were later uplifted along late-rift normal faults, as in the Rincon Hills north of the town, or were exposed by incision of the Rio Grande and its tributaries. During the latter phase of rifting, beginning in the latest Miocene or Pliocene Epochs, new basins and uplifts developed and the ancestral Rio Grande entered the region. Sedimentary rocks deposited in the late-rift basins are also well exposed on this trip.

Excellent exposures of early rift sedimentary rocks are accessible by taking exit 35—Rincon from I-25. Turn south toward the village of Rincon. Go through the village and at a bend in the paved road, just before the railroad crossing, turn right onto a gravel road. A few yards after turning, turn right again, heading toward the red badlands in the foreground. Follow the unpaved road for a hundred yards and park.

The red rocks of the badlands are part of the late Miocene Rincon Valley Formation (Figures 23 and 46a). The red, slope-forming beds are shale containing scattered crystals of the mineral gypsum, which can be easily unearthed with a hammer. Interbedded with the shale are thin, more resistant beds composed of large, shiny crystals of gypsum. Deposition of these rocks took place in a large, ephemeral lake called a playa lake (Figure 46b). The lake probably only had standing water during the rainy season and was dry most of the year. As it dried out, the lake water became saline enough to spontaneously precipitate gypsum.

Retrace the route through the village of Rincon and turn south on I-25 toward Las Cruces. Three miles farther, take exit 32—Upham, turning left after exiting and going under the interstate overpass. After crossing the cattle guard, the road becomes gravel.

One mile from the interstate interchange, go left at a Y-intersection and after 0.2 miles park before the road descends into the canyon.

The gray, red, and white rocks and sediment exposed below the mesa belong to the Pliocene-early Pleistocene Camp Rice Formation and were deposited by the ancestral Rio Grande (Figures 23 and 46c). The gray beds consist of pebbly sandstone that was deposited in the river channels. Most of the sand has not been cemented into a solid sandstone. Red and brown shales and siltstones represent floodplain deposits, many of which contain white nodules of caliche soil. The prominent white bed that can be seen in the distance, to the north, may be a caliche soil or a hot springs deposit. The interbedding of channel sand and floodplain shale indicates that the ancestral Rio Grande moved back and forth across this area numerous times. Indeed, the ancestral Rio Grande traversed a much greater area than the modern Rio Grande, as is discussed in more detail in Field Trip 22. If you look closely in the loose sand, you may find fossil bones and teeth, although they are rare.

The other important geologic feature to be seen at this stop is the ancient soil that developed on the La Mesa surface, which is the flat surface you drove over to reach this spot. This surface represents the top of the Camp Rice Formation. It was the uppermost level of deposition, forming just prior to the beginning of downcutting by the river about 780,000 years ago. When the river and its tributaries began to entrench the basins, ultimately incising to a level about three hundred feet lower, the La Mesa surface was abandoned and a thick, white caliche soil developed (Figure 46d). This soil is very well exposed near the parking spot. Close examination reveals a lower part that is massive, white caliche, capped by a foot or less of finely laminated caliche. The laminar cap formed as rainwater filtered through the upper part of the soil, which was eroded away here, but could not penetrate the massive caliche below. The soil water ponded and precipitated the individual layers of the laminar cap. The Las Cruces area is famous among geologists and soil scientists for exposures of well-developed caliche soils and for the research that elucidated their origin.

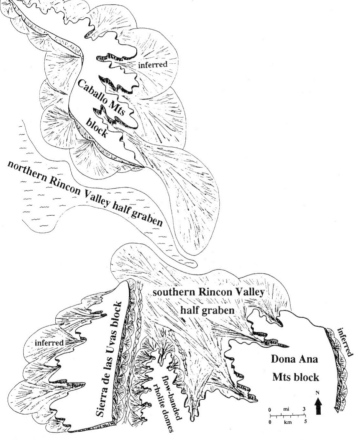

Figure 46. Rincon.
A. Badlands at stop 1 of the Miocene Rincon Valley Formation, composed of red shale and ledges of gypsum deposited in a playa lake.
B. Paleogeographic reconstruction of the area near Hatch, New Mexico, during deposition of the late Miocene Rincon Valley Formation, taken from Mack, G. H., Seager, W. R., and Kieling, J., 1994, Sedimentary Geology, v. 92, p. 79–96. Dashed lines in center of figure represent the ephemeral lake in which were deposited the red shale and gypsum of stop 1.
C. View looking north from the parking spot of stop 2 at exposures of the Pliocene-early Pleistocene Camp Rice Formation, deposited by the ancestral Rio Grande.
D. Mature caliche soil exposed at stop 2.

Field Trip 21:
Grama, NM

Main Theme: Late Tertiary Rio Grande Rift

Access/Difficulty: Natural outcrop reached by offroad vehicle or by car and a moderately difficult three-mile-roundtrip hike on a dirt road

Exposed near the Grama railroad siding is a foot-thick layer of volcanic ash derived from a caldera near Bishop, California. This massive eruption approximately 750,000 years ago sent fine ash over most of the western United States. The ash at Grama has been correlated with the Bishop tuff by its chemical properties and by radioactive dating.

To reach Grama take the Upham exit (exit 32) of I-25. If northbound, turn right; if southbound, turn left and go under the overpass. After crossing the cattle guard the road becomes gravel. One mile from the interstate interchange, go right at a Y-junction, and then immediately right again at another Y-junction, following CR E-72. About 4.4 miles farther, cross a cattle guard and go left at the Y-junction. After 2.2 miles the road crosses a deep arroyo; 0.1 miles after crossing the arroyo, turn left and go toward the windmill. Go through the gate, leaving it as you found it (if closed, close it after you pass; if open, leave it open). It is recommended that those in cars park near the gate and walk the remaining 1.5 miles to the outcrop. The road from here is generally in good shape, but in many places it is sandy and rutted and cars can easily get stuck.

Two-tenths of a mile from the gate, the road crosses an arroyo; 0.5 miles after crossing the arroyo, turn left, following the sign to Grama. The Grama gate appears after 0.8 miles. Walk through the gate, heading due west, and cross the railroad tracks. Be careful, because trains frequently use these tracks and will not blow their horns before approaching this spot. After crossing the tracks, continue walking due west for about thirty yards, until encountering a deep arroyo. The Bishop ash is exposed along the walls of the arroyo, about eight feet below the top of the exposure (Figure 47a, b). There are several places where trails lead down into the arroyo. The ash is white and about a foot thick; dig around a little with

your hammer to see it better. Also exposed here are siltstones and red shales with caliche nodules. Because it is so easily eroded after deposition, ash like this is very rare in the rock record.

Figure 47. Grama. A., B. Exposures of the Pleistocene Bishop ash (between arrows) in the steep arroyo near the Grama railroad siding.

Field Trip 22:
Box Canyon, near Las Cruces, NM

Main Theme: Late Tertiary Rio Grande Rift

Secondary Theme: Early and Mid Tertiary Volcanism

Access/Difficulty: Natural outcrops reached by car and moderately difficult hike, two miles roundtrip, along road and arroyo

Box Canyon in the southern part of the Robledo Mountains has excellent exposures of river deposits of the ancestral Rio Grande. Also present are early and middle Tertiary volcanic rocks. To reach the canyon, take I-10 to exit 135 (U.S. 70). After exiting, cross the dual lanes of Highway 70, following the signs to the Las Cruces Airport. About 1.4 miles after crossing Highway 70, turn right onto a gravel road and immediately cross a cattle guard. A half-mile farther the road bends to the right. Three miles farther the road descends steeply into Box Canyon; those in cars should park here, although offroad vehicles may proceed into the canyon. At the bottom of the canyon is a sign that reads "Box Canyon Wildlife Habitat." Follow the road to the right, going down the major arroyo. On either side of the arroyo, as well as adjacent to the road as it descends into the canyon, are excellent outcrops of the Camp Rice Formation (Figure 48a). Take an hour or so and wander around the cliffs and make observations on these rocks.

The Camp Rice Formation at Box Canyon is Pliocene in age, having been deposited between about 3.5 and 2.5 million years ago. The alternating beds of white or tan pebbly sandstone and red shale were deposited by the ancestral Rio Grande. The sandstones, which are superbly crossbedded (Figure 48b), were deposited in the river channels, whereas the red shale represents deposition on the floodplain. Despite its young age, many Camp Rice sandstones at Box Canyon are well cemented and stand as continuous cliffs. Some of the calcite cement was concentrated into spheres, which weather out as softball-sized, round cobbles that can be found on the outcrop and in the arroyo (Figure 48c). Sedimentation at this site ended when the area was uplifted along the East Robledo fault, forcing the river onto the downthrown block (hanging wall) near Las Cruces.

Figure 48. Box Canyon. A. River deposits of the Pliocene-early Pleistocene Camp Rice Formation consisting of light-colored channel sandstones and darker floodplain silts and mudstones. B. Crossbedded sandstone of the Camp Rice Formation, deposited as subaqueous dunes by the ancestral Rio Grande. C. Softball-sized concretions in crossbedded sandstone of the Camp Rice Formation, formed when the loose sand was cemented into solid rock during burial. D. *(on following page)*

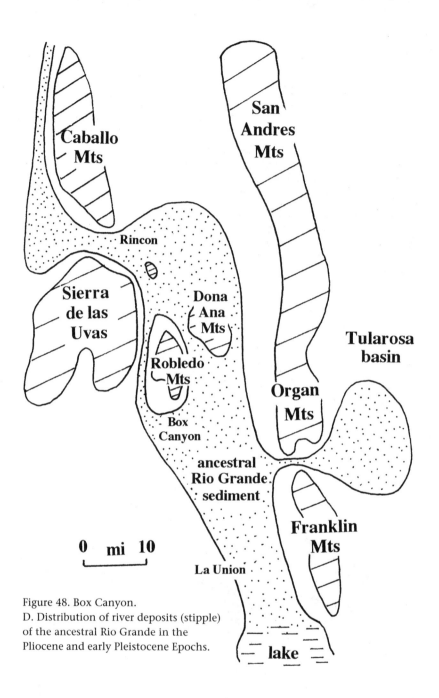

Figure 48. Box Canyon.
D. Distribution of river deposits (stipple) of the ancestral Rio Grande in the Pliocene and early Pleistocene Epochs.

During its 3.2-million-year history, from roughly 4.0 to 0.78 million years ago, the ancestral Rio Grande traversed a very large alluvial plain, an area much larger than that occupied by the modern river (Figure 48d). The ancestral Rio Grande at times flowed in areas that are many miles from the present river, and at least once crossed Fillmore Pass between the Organ and Franklin Mountains, spilling into the Tularosa basin. Near El Paso, the ancestral Rio Grande emptied into a shallow lake (Figure 48d). The ancestral Rio Grande was probably not dissimilar to the modern river, although it may have been a little larger. Then as now, the river was broad and shallow.

Continuing down the major arroyo toward Picacho Peak, there are outcrops of an unconformity between horizontal beds of the Camp Rice Formation above and gently tilted purple and green beds of the Eocene Palm Park Formation below. The Palm Park Formation consists of volcanic andesite deposited as cobbly and bouldery lahars, sandstone and shale, representing volcanic debris reworked by sedimentary processes, and a few resistant beds of lava flows. Picacho Peak, the high, pointed peak to the east of the canyon, is composed of white to red flow-banded rhyolite of Oligocene age.

Glossary of Terms

absolute geologic time. Determining the age of rocks in millions of years before the present by using radioactive decay.

andesite. A gray, green, or purple volcanic rock composed primarily but not exclusively of minerals too small to be seen with the unaided eye.

anticline. A form of deformation in which the rock layers have been bent upward.

arc, volcanic. The arcuate trail of mostly andesite and rhyolite volcanoes on the overriding plate above a subduction zone.

ash. Microscopic pieces of rock, mineral, and glass blown out of a volcano.

ash fall. Volcanic particles blown into the air by an explosive eruption and later falling on the surrounding countryside.

ash flow. A mixture of hot gas and particles that flows down the flank of an explosive volcano.

asthenosphere. The zone of partially molten mantle directly beneath the lithospheric plates.

basalt. A dark volcanic rock composed primarily, but not exclusively, of minerals too small to be seen with the unaided eye.

basement. The crystalline core of a continent, composed of granitic and metamorphic rocks.

bed/bedding. Primary layering of sedimentary and volcanic rocks produced at their time of formation.

caldera. A large, circular to elliptical depression created by the collapse of a volcanic vent.

caliche. A type of soil common in arid regions composed of whitish, microscopic crystals of the mineral calcite.

chemical sedimentary rock. Sedimentary rock produced by organic or inorganic precipitation of minerals from lake water or sea water.

clay. A class of microscopic minerals produced by chemical weathering.

columnar joints. Near vertical columns formed by cooling and shrinking of volcanic rocks.

compression. Stress in the earth in which rocks are squeezed together.

conglomerate. A detrital sedimentary rock composed of gravel-sized particles.

continental accretion. The process whereby continents grow outward through time by a series of collisional events.

convergence. When tectonic plates move toward each other.

crossbeds. A type of sedimentary structure composed of inclined layers formed when moderate-velocity wind or water currents transport and deposit silt, sand, fossils, or oolites as dunes.

cross cutting relationship. When magma cuts across pre-existing rocks and subsequently crystallizes into an igneous rock; the igneous rock is younger than the rocks it cross cuts.

deformation. Bending and breaking of rocks as a result of natural forces acting in the earth.

desiccation cracks. A sedimentary structure produced by the drying and cracking of sediment.

detrital sedimentary rock. A type of sedimentary rock composed of particles produced by weathering and transported and deposited by water, wind, and gravity.

diorite. A gray plutonic igneous rock composed of minerals visible with the unaided eye.

dip. The direction and angle of tilt of deformed rocks.

divergence. When tectonic plates move away from each other.

dolomite/dolostone. A chemical sedimentary rock composed of the mineral dolomite; represents a limestone altered during burial.

element. One of the fundamental units of matter and the building blocks of minerals.

eon. The largest subdivision of geologic time; group of eras.

epoch. The smallest subdivision of geologic time; subdivisions of a period.

era. Group of geologic periods.

evaporite. A class of chemical sedimentary rock that forms by inorganic precipitation of minerals, such as halite or gypsum, from hypersaline lake water or sea water.

extension. Stress in the earth in which the rocks are pulled apart.

fault. A break in the earth in which the rocks move past one another.

foliation. The alignment or segregation of minerals in metamorphic rocks produced by a directed stress.

footwall. The block of rock that would be underfoot if walking on a fault plane.

formation. A subdivision of rocks in a given region for the purpose of geologic mapping; geographic names are applied to formations.

gabbro. A dark plutonic igneous rock composed of minerals visible with the unaided eye.

geologic time scale. Subdivision of earth history into Eons, Eras, Periods, and Epochs using fossils and radioactive decay.

glass, volcanic. Lava that solidifies so quickly that minerals cannot form; includes pumice and obsidian.

gneiss. A metamorphic rock containing minerals visible with the unaided eye and segregated into light and dark bands.

granite. A white, tan, pink, or red plutonic igneous rock composed of minerals visible with the unaided eye.

hanging wall. The block of rock that would be overhead if walking on a fault plane.

hot spot. Intraplate volcanism.

igneous rock. A class of rocks in which minerals crystallize from a molten magma; includes volcanic and plutonic varieties.

lahar. Volcanic mudflow.

limestone. A type of chemical sedimentary rock composed of the mineral calcite.

lithospheric plates. Thin, rigid plates that move across the earth's surface.

marble. A metamorphic rock representing a recrystallized limestone or dolomite.

metamorphic rock. A class of rocks formed by recrystallization of pre-existing rocks under conditions of high temperature and pressure.

metaquartzite. A metamorphic rock representing recrystallized sandstone.

minerals. The natural combination of elements that make up rocks.

normal fault. A fault produced by extension, in which the footwall is moved upward with respect to the hanging wall.

oolites. Inorganically precipitated, sand-sized spheres of calcite found in some limestones.

paleogeographic map. Interpretation of the salient geographic features of a region during a particular interval of geologic time.

paleomagnetism. Alignment of iron atoms in the direction of the earth's magnetic field during the time a rock forms.

period. The basic unit of the geologic time scale.

plate tectonics. A theory stating that the outer portion of the earth is divided into rigid plates that are in motion across the earth's surface.

plutonic rock. A variety of igneous rock formed when molten magma crystallizes below the earth's surface; composed of minerals visible with the unaided eye.

pumice. A light-colored, vesicular volcanic glass.

pyroclastic. Mineral, rock, and glass ranging in size from dust to houses blown from a volcano.

relative geologic time. Comparing the age of rocks that are in physical contact with each other and/or by the fossils they contain.

rhyolite. A white, tan, pink, or red volcanic rock composed primarily, but not exclusively, of minerals too small to be seen with the unaided eye.

rift. Continental crust that is being pulled apart.

ripple marks. A sedimentary structure consisting of a regular pattern of inch-scale undulations produced when low-velocity wind and water currents transport and deposit silt, sand, fossils, or oolites.

rocks. The natural combination of minerals that make up the solid earth.

sandstone. A detrital sedimentary rock composed of sand-sized particles.

schist. A metamorphic rock containing shiny minerals of mica aligned in a plane and large enough to be seen with the unaided eye.

sedimentary rock. A class of rocks composed of material produced by weathering and deposited at the earth's surface.

shale. A detrital sedimentary rock composed primarily of clay minerals.

siltstone. A detrital sedimentary rock composed of silt-sized particles.

slate. A metamorphic rock composed of minerals too small to be seen with the unaided eye; splits into flat sheets because the microscopic minerals are aligned.

spheroidal weathering. The tendency for igneous rocks to weather into spherical bodies.

spreading ridge. Linear chain of oceanic basalt volcanoes formed where tectonic plates move away from each other.

strike-slip fault. A break in the earth where the rocks slide horizontally past one another.

stromatolites. Fossil colony of cyanobacteria (blue green algae) resembling cabbage heads.

structural geology. Branch of geology dealing with rock deformation.

subduction. The process in which one tectonic plate moves beneath another.

syncline. A form of deformation in which the rock layers have been bent downward.

thrust fault. A fault produced by compression in which the hanging wall is moved up with respect to the footwall.

trace fossil. Fossil tracks, trails, and burrows.

tuff. A volcanic rock composed of pyroclastic material.

unconformity. A surface between two rocks formed by a long period of erosion.

vesicle. A hole in a volcanic rock formed when gas escapes the lava.

volcanic rock. A type of igneous rock formed when molten lava solidifies at the earth's surface; composed primarily, but not exclusively, of minerals too small to be seen with the unaided eye.

weathering. Natural disintegration of rocks at the earth's surface.

xenolith. A fragment of rock broken off by moving magma and surrounded by the igneous rock when it solidifies.

Further Reading

There are a number of books at the first-year college level that deal with the basic principles of geology. These are available in college or university libraries, and perhaps in some public libraries. Some notable examples covering physical geology are *The Dynamic Earth*, by Brian J. Skinner and Stephen C. Porter (John Wiley and Sons, 1992) and *Geology*, by Stanley Chernicoff and Ramesh Venkatakrishnan (Worth Publishers, 1995). Among the best historical geology books, which concentrate on earth history and the fossil record, are *Earth and Life through Time*, by Steven M. Stanley (W. H. Freeman and Co., 1989) and *Evolution of the Earth*, by Robert H. Dott, Jr., and Donald R. Prothero (McGraw-Hill, 1994).

Also valuable in the context of field geology are geologic maps. Three geologic maps covering the area of southwestern New Mexico discussed in this book are published by, and can be purchased from, the New Mexico Bureau of Mines and Mineral Resources, Socorro, NM 87801. These maps are New Mexico Bureau of Mines and Mineral Resources Geologic Maps 53 (W. R. Seager et al., 1982), 57 (W. R. Seager et al., 1987), and 60 (W. R. Seager, 1995). A geologic map that covers the entire state of New Mexico is entitled *New Mexico Highway Geologic Map*, which is published by the New Mexico Geological Society and can be purchased from the New Mexico Bureau of Mines and Mineral Resources. An appropriate geologic map of west Texas is *Geologic Atlas of Texas, Van Horn-El Paso Sheet*, which can be purchased from the Bureau of Economic Geology, the University of Texas at Austin, Austin, TX 78712.

Index

Photographs are indicated by (ph); illustrations by (illus); maps by (map).

Abo Formation, 48 , 49, 50, 94(ph), 96, 97
absolute geologic time, 21, 22
accreted terranes, 38(illus)
accretion, continental, 39
Achenback Park, 61
Acklin Hill Road, 79
Aguirre Springs State Park, 137
Alamogordo, NM, 93
Aleutian Islands, 29
algae, 12, 40, 110; fossil, 93; phylloid, 93
alluvium, Quaternary, 114
Alpine, TX, 27
Anapra Formation, 53, 113, 114, 115
Ancestral Rocky Mountains, 46–50, 101; field trip viewing of Permo-Pennsylvanian, 93–97, 98–102, 103–5, 106–11; late Paleozoic paleogeography of, 46(illus); theories of origin of, 49–50
andesite, 9, 10, 19(ph), 29, 62, 99, 103, 126(ph), 135, 145, 147; in lahars, 161
andesite dikes, 95(ph), 97, 125, 128
Andes Mountains, 29
angiosperms, 58
Animas Mountains, 140, 141
anticlines, 26, 111
Apache Canyon (Caballo Mountains), 146–51, 146(ph), 149(ph)
Apache Hill, 91, 92(ph)
Appalachian Mountains, 26
Archean Era, 21
Arches National Park, 51
Arctic Ocean, 55
arc volcanism, 29, 30(illus), 59
Artesia Group, 49, 110
ash, 7, 157; and calderas, 65
ash deposits, 7
ash-fall deposits, 7, 142(ph), 144, 145
ash falls, 131
ash flows, 7, 8, 125, 126(ph), 132; in outcrop, 141
ash-flow tuffs, 87, 125, 131(ph), 131, 132, 136, 142(ph)
asthenosphere, 28, 29

back reef, 110
badlands, 152, 155(ph)
Bandelier National Monument, 8
Banff National Park, 55

barite, 89, 90, 124
basal bed, 126(ph), 127, 128,
basalt, 9, 10, 79, 125, 127(ph), 145; with cooling joints, 128; and rhyolite, 64
basalt volcanoes, 29
basement, 40
Basin and Range, 68(illus), 71
beds or bedding, 12
Beeman Formation, 49, 96
Bell Top Formation, 61, 105
bentonites, 57
Big Hatchet Mountains, 52
Bighorn Mountains, 60
biotite, 81, 116, 137, 139, 140, 145; equant, 75
Bisbee basin, 51, 52(illus)
Bishop's Cap, 87–90, 88
Bishop ash, 157(ph)
Black Hills, 60
Black Range, 129, 140–45
Bliss Formation, 43, 45, 79, 80(ph), 83(ph), 84
block faulting, 144
block rotation, 87
bolide, 63
Boquillas Formation, 53
Box Canyon (Robledo Mountains), 158–61, 159(ph), 160(map)
brachiopods, 81, 85, 91, 105
brachiopod shells, 90
bryozoans, 81, 90, 91
Buda Formation, 53, 114, 115(ph), 115–16
Buda Limestone, 116
Burro uplift, 52(illus)
burrows, 13, 85, 86(ph), 118, 119(ph), 120

Caballo dam, 27
Caballo fault, 27
Caballo Mountains, 148(ph), 160(map); and Apache Canyon, 146–51
Caballo Mountains block, 154(map)
Caballo Mountains uplift, 117
calcite, 12, 15, 91; microcrystalline, 89; nodules of, 94, 101
calderas, 8, 64–66; development of, 144; field trip viewing of Mid Tertiary, 136–39, 140; location of Oligocene, 65(map); near Bishop, CA, 156
caliche, 50, 101, 124, 155(ph); clasts of, 94; nodules of, 153, 157
caliche paleosol, 94(ph)

caliche soil formation, 70
Camp Rice Formation, 61, 98, 99, 100, 123(ph), 124, 153, 155(ph), 158, 159(ph), 161
Campus Andesite, 116
Canyonlands National Park, 51
Carlsbad Caverns, 48, 109(ph); morphology of, 111
Carrizozo, NM, 69
Cascade Arc, 68(illus)
Cascade Range, 29
Castile Formation, 49, 108(ph)
Castner Formation, 75, 77(ph)
Cedar Hills, NM, 129–32
Cedar Hills Vent Zone, 65(map), 129, 130(ph), 132
cementation, 11, 13
cephalopods, 91
Cerro de Cristo Rey, 52, 112–16, 114(map), 115(ph)
chert, 85, 90
Chihuahua Trough, 51–54, 52(illus), 112, 113; field trip viewing of Early Cretaceous, 112; fossil record of, 54
Chinle Formation, 51
Chino open pit mine, 80(ph), 81
Ciudad Juárez, 85
clams, 91, 115, 116
clam shells, 101, 105
clasts, 94, 107, 108(ph)
clay, 101
clay minerals, 11
climate, 37, 63
coal deposits, 57
Cobre fault block, 79
Cobre Uplift, 79–81, 80(ph)
colluvium, 79
color of rocks, 10
columnar joints, 8
compaction, 11, 13, 121
compression, 24, 26
compressional mountain building, 59
concretions, 159(ph)
conglomerate rock, 11
conglomerates, 94, 97, 100(ph), 107, 108(ph), 132, 140, 147, 149(ph), 151
conifers, 50
continental crust, 32
convection cells, 28
convergence, 29
converging plates, 30(illus)
cooling fractures, 137, 145
cooling joints, 127(ph), 131(ph), 138(ph), 139, 141, 142(ph)
corals, 90, 116
couplets, 110
Cox Ranch Formation, 61, 136, 137

cracks, 8
Cretaceous paleogeography, Early, 52(illus)
Cretaceous rocks, 113, 121
Crevasse Canyon Formation, 53, 117, 121
crinoid heads, 91
crinoid plates, 89, 90, 91, 105
crinoids, 81
crossbedded sand, 148(ph)
crossbedding, 118, 158
crossbeds, 13, 14(ph), 83(ph), 96, 101, 113
cross cutting relationship, 18, 19(ph)
crustal extension, 67, 71
crystalline basement, 75
crystals, 136
Cueva ash-flow tuff, 137, 138(ph)
Cueva Formation, 61, 87, 136
Cueva Rock, 138(ph)
Cueva Rock trail, 137
cyanobacteria, 40

Dakota Formation, 117, 118
dating of rocks, 18–21; problems with using fossils in, 20–21
D-Cross Formation, 53, 117, 120
deformation, 24, 147
Delaware basin, 47, 106, 110
Delaware Mountains Group, 49, 106, 108(ph), 110
Del Norte Formation, 53
Del Rio Formation, 53, 113, 115
Deming, NM, 133
deposition, 107, 118
depositional environment, 13
desiccation cracks, 14(ph), 15, 75, 101
detrital sedimentary rocks, 11
detritus, 11, 12, 68
dikes, 125, 127(ph), 128
Dinosaur National Monument, 51
dinosaurs, 58; demise of, 62
diorite, 9, 10, 29, 81, 116; Muleros, 115(ph)
dip, 24
divergence, 29
diverging plates, 31(illus)
dolomite, 12, 15, 16, 79, 80, 85, 94, 101, 105, 106, 109(ph), 110–11, 149(ph); back reef, 109(ph)
dolomitization, 12
dolostone, 12
Doña Ana Mountains, 48, 98, 100(ph), 154(map), 160(map)
downcutting, 70–71
Dripping Springs State Park, 136–37
Dripping Springs trail, 137, 138(ph)

Dry Canyon, 93–102
dunes, 13, 15
dynamism of earth, 3

earth, age of, 17
earthquakes, 33
earthquake waves, 25
East African rift, 32
East Potrillo Mountains, 51
East Robledo fault, 103, 104(ph), 158
echinoid plates, 105
echinoids, 101, 116
echinoid spine, 100(ph)
El Capitan, 106, 108(ph)
elements, 5; combination of, 5
Elephant Butte dam, 27
El Paso, TX, 4(map), 75, 82, 85–86
El Paso Formation, 45, 80, 83(ph), 84, 85, 86(ph)
Emory caldera, 64, 81, 138, 140, 142(ph); development of, 143(illus)
Emory Pass, 141
Emory Pass Vista Point, 144, 145
Entrada Formation, 51
eons, 20, 20
epidote, 135
epochs, 20
equator, former location of, 42
equipment needed for field geology, 2
eras, 20
etiquette, geologic, 74
evaporation, 12
evaporites, 12
extension, 24
extinction, mass, 54, 62–63

Farallon plate, 55, 56, 59, 66, 99, 124
Faulkner Canyon, 125, 126(ph)
faults, 24; explanation of working of, 27; rotation of, 26; types of, 25–26
feldspar, 15, 76(ph), 78, 81, 116, 128, 129, 131, 135, 136, 137, 139, 140, 141; enveloped by obsidian, 129; in granite, 139
field trips, difficulty rating of, 73
Fillmore Pass, 161
Finley Formation, 53
Florida Mountains, 133–35, 134(ph)
flow banding, 66, 131, 132
fluorite, 89
folds, 26, 107, 110; syndepositional, 108(ph)
foliation, 15, 16
footwall, 26
foreland basin, 55
formations: definition of, 35; naming of, 35

fossil algae, 93
fossil collecting, 91, 105, 117
fossil corals, 85, 86(ph)
fossiliferous beds, 91
fossil record, 66, 72
fossils, 101, 103, 107, 110, 115; as key to age of rocks, 18; fusulinid, 95(ph); Paleozoic invertebrate, 44(ph)
Foster Canyon, 132
Franklin Mountains, 26, 75, 83(ph), 85, 160(map), 161; stratigraphy of, 82(illus)
Fresnal fault, 94(ph), 95(ph), 96, 97
Front Range, 60
Fusselman Formation, 45, 80, 87, 88, 89
fusulinid fossils, 95(ph), 97

gabbro, 9, 10, 29, 79
Gallup Formation, 53, 117, 119(ph), 120, 121
garnet, 15, 77
gastropods, 85, 100(ph), 115(ph), 116
geodes, 133
Geological Sciences Department (New Mexico State Univ.), 66
geologic time scale, 18, 20
Gila Conglomerate, 145
Glacier National Park, 55
glauconite, 79
gneiss, 15, 16, 123(illus)
Gobbler Formation, 49, 94(ph), 96
Gondwanaland, 48
Grama, NM, 156–58, 157(ph)
granite, 9, 10, 29, 32, 75(ph), 123(illus), 139; cross cutting metamorphic rock, 19(ph); intrusive, 136; Precambrian, 124; weathering of, 138(ph)
gravels, Pleistocene, 126(ph), 128
Great Salt Lake, 12
Guadalupe Mountains, 48, 106–11, 108(ph)
Gulf of Mexico, 55
gypsum, 12, 48, 108(ph), 110, 152, 154(ph)

halite, 12, 48
hanging wall, 26
Hatch, NM, paleogeographic reconstruction of area near, 154(map)
Hawaiian Islands, 33
Hayner Ranch Formation, 61, 147, 149(ph), 150–51
Helms Formation, 45, 87, 88, 90
Herculaneum, 8
Hillsboro, NM, 91, 141
Himalaya Mountains, 32

Holder Formation, 49, 93, 94(ph), 97
hornblende, 76(ph), 116, 128, 135, 139
horn corals, 91
hot spots, 33
Hueco Formation, 49, 99, 100(ph), 101, 103, 104(ph), 105
hydrocarbons, 121

igneous intrusion, field trip viewing of Early Tertiary, 112
igneous rocks, 6–10, 80, 112, 116; color of, 10; plutonic, 9; volcanic, 9
intraclasts, 86

Japanese islands, 29
jasper, 90, 133
Jasper National Park, 55
Jemez caldera, 8, 70
Jemez Volcanic Field, 70
Jesus, statue of, 112, 115

Kent Hall Museum (New Mexico State Univ.), 58, 72
Kilbourne Hole, 69
Kingston, NM, 144
Kneeling Nun Formation, 61, 81, 140
Kneeling Nun tuff, 142(ph), 143(ph), 144, 145
Kobi, Japan, 29
Laborcita Formation, 49, 50, 93, 94(ph), 96

lahars, 7, 62, 99, 100(ph), 125, 126(ph), 128, 135, 144, 161
Lake Lucero, 12
Lake Valley, NM, 91–92, 92(ph)
Lake Valley Formation, 45, 87, 88, 89, 91, 92(ph)
La Luz Canyon, 93–97
La Mesa surface, 153
laminae, 113
laminar cap, 153
Lanoria Formation, 76(ph), 78
Laramide, 60
Laramide deformation, field trip viewing of Early Tertiary, 122–24; 146
Laramide fault, 122, 123(ph)
Laramide Volcanism, field trip viewing of, 98–102; field trip viewing of Early Tertiary, 125
Laramie, WY, 60
Las Cruces, NM, 87, 103
La Tuna Formation, 87, 88, 90
lava flows, 7, 125, 126(ph), 128, 130(ph), 133, 135, 161; basalt, 141, 142(ph)

layering, 131
limestone, 80, 81, 85, 91, 92(ph), 98, 99, 101, 103, 104(ph), 105, 107, 108(ph), 110, 112, 144–45, 149(ph); fossiliferous, 12, 101, 116; metamorphosed to marble, 16, 144; micrite, 12; oolite, 90; Paleozoic, 94, 124, 147; reef, 106; shells as source of, 12; silicified Paleozoic, 123(illus)
limestone cliffs, 106
Lingula, 88, 89
lithic tuff, 137
lithospheric plates, 28
Little Florida Mountains, 133, 134(ph)
Little Hatchet Mountains, 52
Lobo Formation, 61
Los Alamos, NM, 70
Love Ranch Formation, 61, 147, 148(ph), 150; outcrop of, 148(ph)
Lucero Arroyo, 98–102, 100(ph)

Magdalena Group, 49, 87, 88, 147, 148(ph), 149(ph), 150
magma, 6; basaltic, 29
magnetic force lines, 36(illus)
Mancos Formation, 53, 117, 118, 119(ph)
maps, geologic, 73
marble, 15, 16, 75(ph), 77, 144
marine environments, 13
McKelligon Canyon, 83(ph)
McKelligon Canyon Park, 82–84
McKittrick Canyon, 106, 107
McRae Formation, 53
Mescal Canyon, 117–21, 119(ph); outcrop in, 119(ph)
Mesilla basin, 98, 103
Mesilla Valley Formation, 53, 113, 114
metamorphic rocks, 15–16, 32, 40
metaquartzite, 15, 16, 75, 78, 97
micas, 15, 16, 136
Mimbres-Sarten fault, 79
minerals, 5, 15; combinations of, 6; precipitation of, 12; transportation of, 11
mines, 89
moats, 143(illus), 144
Modoc fault, 137
Mollusks, 54
Montoya Formation, 45, 80, 83(ph), 84, 85, 86(ph)
Morrison Formation, 51
Mount Cristo Rey, 112, 114, 115(ph), 116
Mount Withington caldera, 150
mudstone, 159(ph)
muggings, 112

INDEX 173

Muleros Andesite, 116
Muleros diorite, 115(ph)
Muleros Formation, 53
Murchison Park (El Paso, TX), 85–86, 86(ph)

Navajo Formation, 51
New Guinea, 39
New Madrid, MO, 33
normal faults, 25, 87, 121, 122(ph), 124, 137, 140, 147, 152; East Robledo, 103
nuclear waste storage, 48
Nutt, NM, 91

obsidian, 65, 129, 130(ph), 131, 133
obsidian flow, 132
Oligocene Epoch, 64
oolites, 12
Orejon Formation, 61
Organ Mountains, 98, 100(ph), 129, 160(map), 161; granite of, 138(ph), 139
Organ Mountains caldera, 64, 87, 136, 138(ph)
Organ Mountains fault, 27
Orogrande basin, 47, 101, 105
Ouachita Mountains, 26
oysters, 115, 116, 121; fossils of, 119(ph)

Padre Island, 84
Painted Desert National Park, 51
paleogeographic maps, 35–37
paleogeography, Late Cretaceous, 56(illus)
paleogeography, Late Cretaceous/Early Tertiary, 60
paleolatitude, 36
paleomagnetism, 37
paleosol, 94, 121
Paleozoic age, 43
Paleozoic sedimentation, field trip viewing of Early-Mid, 79–92
Palm Park Formation, 61, 99, 100, 103, 125, 126(ph), 127(ph), 128, 147, 161
Palomas basin, 140, 151
Palomas Formation, 61, 147, 148(ph)
Pangaea, 50
parallel motion, 29, 32
Pedernal Mountains, 47, 93, 94, 97
Pedernal Uplift, 96
Peloncillo Mountains, 52
Percha Formation, 43, 45, 80, 81, 87, 88, 89, 144
periods, 18, 20
perlites, 133
Permian reef complex, 48
Petrified Forest National Park, 51

phreatomagmatic eruption, 69
phylloid algae, 93
phylloid algae reef, 94(ph)
Picacho Peak, 161
pisolites, 111
plate configurations, changes in, 33
plates, three ways of interaction of, 29
plate tectonics, 28–33
playa lake, 152, 154(ph)
plutonic rocks, 9, 10, 134; relative age of, 18
Pompeii, 8
Precambrian basement, 75, 79, 84
precautions for field geology, 2
Proterozoic Era, 21
pumice, 7, 70, 137, 141, 142(ph), 145, 150
pyroclastic material, 6

quarries, 113, 115(ph)
quartz, 15, 76(ph), 78, 81, 89, 124, 131, 136, 137, 139, 140, 145; enveloped by obsidian, 131

radioactive decay, 21
radiometric dating, 21
Rancheria Formation, 45, 87, 88, 90
recrystallization, 15, 110
Red Bluff Formation, 75
Red Bluff Granite, 78, 84
Red Hills fault, 147, 148(ph), 151
Red Sea, 12
reef core, 108(ph)
relative geologic time, 17, 21
rhyolite, 9, 10, 29, 64, 65, 78, 105, 133, 142(ph), 144; flow-banded, 131, 134(ph), 142(ph), 161; Thunderbird, 76(ph), 78
rhyolite ash-flow tuff, 140
rhyolite calderas, 129
rhyolite domes, 65, 130(ph), 131; flow-banded, 129, 142
rhyolite tuffs, extrusive, 136
rhyolite volcanic sand, 129(ph)
rifting, 67
Rincon, NM, 153–55, 154–55(ph)
Rincon Valley Formation, 61, 152, 154–55(ph)
Rio Grande rift, 67–72, 68(illus), 82, 87, 98, 103, 133, 140, 150; crustal extension in, 146; deformation in, 147; fault block locations in, 69(map); field trip viewing of Late Tertiary, 122–24, 146–51, 152–55, 156–57, 158–61; fossil record in, 72
Rio Grande River: ancestral, 122, 140, 153, 158, 161; downcutting in, 70–

174 INDEX

71; integration of upper and lower parts, 71
ripple marks, 13, 14(ph), 101, 107
Robledo Mountains, 103–5, 104(ph), 125, 158, 160(map)
Rock Hound State Park, 133, 134(ph)
rocks, 6; dating of, 18–22; marine, 117; movement of, 11; nonmarine, 47, 117; overlying, 22–23; upward movement of, 24; volcaniclastic, 132. *See also specific rock types and rock names*
Rocky Mountains. *See* Ancestral Rocky Mountains
roots, coalified, 121
root traces, 13
Rubio Peak Formation, 61, 133, 134(ph), 144

Sacramento Mountains, 47, 94–95(ph)
safety considerations, 2
Salado Formation, 49
San Andreas fault, 26, 29
San Andres Formation, 49
San Andres Mountains, 160(map)
San Austin Pass, 136–39
San Diego Mountain, 123(ph), 122–24
sandstone, 11, 79, 83(ph), 84, 94, 110, 112, 113, 115(ph), 117, 118, 119(ph), 123(ph), 140, 141, 144, 145, 153, 158, 159(ph), 161; back reef, 108(ph); and coalified roots, 121; crossbedding in, 96; deep-water, 106; interbedded, 151; metamorphosed, 16, 78; turbidite, 107; volcanic, 126(ph), 128
San Juan basin, hydrocarbon reserves of, 57
San Lorenzo, NM, 79, 145
schist, 15, 16
sea level, changes in, 32, 43, 48, 113, 117
sedimentary environments, nonmarine and marine, 13
sedimentary rocks, 10–15, 21, 87, 106, 112, 117, 141, 146, 152; chemical, 11; detrital, 11; layering of, 12–13; Phanerozoic, 21; structures in, 13–15
sedimentary structures, 14(ph)
shale, 11, 80, 88, 93, 94(ph), 97, 101, 105, 106, 107, 108(ph), 112, 113, 115(ph), 117, 118, 119(ph), 141, 142(ph), 144, 147, 149(ph), 153, 154(ph), 158, 161; fossiliferous, 104(ph); with gypsum crystals, 152; and hydrocarbons, 121
shells, 12, 45; ammonite, 58; clam, 105, 118; oyster, 113
Sierra de las Uvas, 154(map), 160(map)

Sierra Juarez, 26, 51
silt, 11, 159(ph)
siltstone, 11, 87, 88, 93, 98, 100(ph), 101, 103, 107, 112, 113, 117, 118, 119(ph); highly fractured, 99
slate, 15, 16, 75(ph), 77
Smeltertown Formation, 53
snails, 91, 101, 105, 116
spheroidal weathering, 9, 76(ph), 126(ph), 128, 145
spines, 105
sponges: calcareous, 110; fossil, 108(ph), 109(ph)
spreading ridges, 29, 31(illus), 32
Spring Canyon State Park, 133, 134(ph)
Squaw Mountain Formation, 61
Squaw Mountain tuff, 137, 138(ph)
stratigraphy: Cenozoic, 61; Cretaceous, 53; Late Paleozoic, 48; Paleozoic, 45
stratovolcanoes, 62, 125, 135
stress, 24, 25
strike-slip fault, 25, 26
stromatolites, 41
structural geology, 24
subduction, 29, 30(illus)
subduction angle, decrease in, 59
Sugarloaf Peak, 138(ph), 139
syncline, 26

Tertiary rocks, 125
Tertiary Volcanism: field trip viewing of Early and Mid, 133–34, 158–61
Texas Bureau of Economic Geology, 107
thrust faults, 25, 26, 60, 62
Thunderbird Formation, 78
Thurman Formation, 61, 149(ph), 150
tidal channels, 100(ph), 101
tidal flats, 110
trace fossils, 13, 14(ph)
tracks, 13
trails, 13
Transcontinential Arch, 42(illus)
Trans-Mountain Road, 75
Tres Hermanos Formation, 53, 117, 118, 119(ph), 120
trilobites, 91
Truth or Consequences, NM, 117
Tuff of Achenback Park, 87
tuffs, 6, 8, 9, 64–65
Tularosa basin, 160(map), 161

Uinta Mountains, 60
unconformity, 22, 43, 79, 84, 85, 99, 100(ph), 126(ph), 128, 142(ph), 147, 149(ph); outcrops of, 161; types of, 23
Uvas basalt, 141, 142(ph)
Uvas Formation, 61, 141

Vado Andesite, 116
Valley of Fires State Park, 69
vent zones. *See* Cedar Hills Vent Zone
vesicles, 6, 7
volcanic arc, 29, 125
volcanic ash, 156
volcanic eruptions, types of, 6
volcanic glass, 6
volcaniclastic rocks, 132
volcanic particles, movement by wind of, 7
volcanic rocks, 6, 9, 10, 98, 125, 128, 133, 134(ph), 135, 140, 141; types of deposition of, 7–8
volcanic sandstone, 135
volcanism, 69; andesitic, 99; field trip viewing of Mid Tertiary, 129
volcanoes: basalt, 29; and spreading ridges, 29

Waste Isolation Pilot Project, 48
weathering, 10–11. *See also* spheroidal weathering
Western Interior Seaway, 55–58, 117, 118; field trip viewing of Late Cretaceous, 117–21; marine life of, 57–58
West Potrillo Mountains, 69
White's City, NM, 106, 110
White Sands Missile Range, 27
Wilderness Park Museum, 75
Wind River Range, 60

xenoliths, 75, 76(ph)

Yellowstone National Park, 7, 33
Yeso Formation, 49, 96
Yosemite National Park, 55

Zion National Park, 51